Why is it dark at night?

Why is it dark at night?
Story of dark night sky paradox

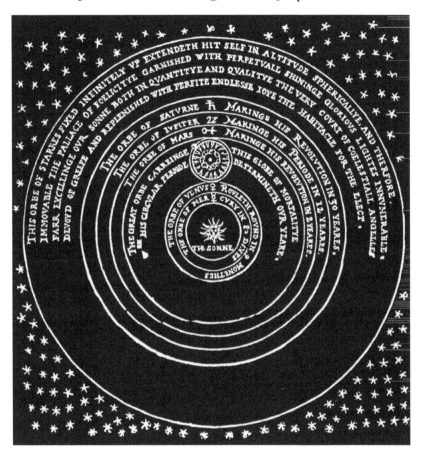

Peter Zamarovský

AuthorHouse™ UK Ltd.
1663 Liberty Drive
Bloomington, IN 47403 USA
www.authorhouse.co.uk
Phone: 0800.197.4150

© 2013 by Peter Zamarovský
(first Czech edition 2011).
Translation © Gerald Turner 2013.
All rights reserved.

No part of this book may be reproduced, stored in a retrieval system, or transmitted by any means without the written permission of the author.

Published by AuthorHouse 11/12/2013

ISBN: 978-1-4918-7880-4 (sc)
ISBN: 978-1-4918-7879-8 (hc)
ISBN: 978-1-4918-7881-1 (e)

Because of the dynamic nature of the Internet, any web addresses or links contained in this book may have changed since publication and may no longer be valid. The views expressed in this work are solely those of the author and do not necessarily reflect the views of the publisher, and the publisher hereby disclaims any responsibility for them.

Contents

Foreword vii
Author's foreword ix
Mysteriously self-evident 1
Billions of alien suns, or why it ought not to be dark at night 3
Why it is dark at night 12
The Universe of the Stoics 25
The Aristotelian-Ptolemaic Cosmos 29
The Renaissance and the return of the Epicurean Universe 33
Paradoxes around the Dark Night Sky Paradox 36
The Stoic and Aristotelian Universes Reincarnated 39
The End of the Eternal Universe 59
Astrophysics arrives on the scene 67
The path to the Big Bang 74
A paradox that resisted solution and the Standard Model 85
Darkness at night and its total cause 92
The players in our story 93
Bibliography 159
About the author 163
Subject Index 165
Index of Personal Names 169

Foreword

"Why is it dark at night?" might seem a fatuous question at first sight. In reality it is an extremely productive question that has been asked from the very beginning of the modern age, not only by astronomers, for whom it is most appropriate, but also by physicists, philosophers, and even poets.

The book you have just opened uses this question as a pretext to relate in the most interesting way the history of human thought from the earliest times to the here and now. The point is that if we want to appreciate the magic power of this ostensibly naïve question we need to discover how it fits into the wider context of the natural sciences and learn something of the faltering steps towards an answer.

In doing so the author also guides us through periods that we regard as the dim and distant past. However, as we start reading these passages we are amazed to discover just how searching were the questions the ancient philosophers asked themselves in spite of their fragmentary knowledge of the universe, and how clairvoyantly they were able to gaze into its mysterious structure. But they thought that the universe was unchanging over time because the fixed stars were really fixed, both as regards their mutual positions, and also their brightness and colour.

It will probably not come as a surprise to the informed reader that as in many other fields the crucial question about the night sky was formulated by one of the first great luminaries of the Renaissance period of astronomy, physics and mathematics, Johannes Kepler, although he didn't come up with a solution; the fact is that he wouldn't have been able to because for one thing, at that time no one had any inkling of stellar distances and geometrical dimensions and for another physics only became a reputable science thanks to such figures as Galileo Galilei, Robert Hooke, Isaac Newton, Blaise Pascal and a host of others.

The author goes on to explain very graphically how this increasingly prickly question was tackled by many great men of science in the 19th century, albeit not very successfully overall. It is bound to come as a surprise that it was not a philosopher, a physicist or an astronomer, but instead the poet Edgar Alan Poe, who hinted at the right answer. A poet's intuition is clearly a powerful instrument, as in the case of our Czech national poet Jan Neruda, whose Songs of the Cosmos brilliantly anticipated a number of astrophysical discoveries by an entire century.

The final chapter of the book brings a denouement as in the best detective stories, and I'm sure that inquiring readers will all relish it. Likewise they will appreciate the freshness of the author's style and the graphic presentation of this slim volume, which contains remarkable and often revealing supplementary information and is studded with little-known quotations from the works of scholars and philosophers from ancient times up to the present day, all demonstrating the erudition and pedagogical talents of Dr. Peter Zamarovský. The author lectures in physics and the philosophy of science and technology at the Electrical Engineering Faculty of the Czech Technical University in Prague, but is also currently chairman of the European Cultural Club, which, for over twenty years, has organised public panel discussions on topical issues related to various fields of science, philosophy and art in the main building of the Academy of Sciences.

The author's wide horizons are also evident in the forty or so brief profiles of the main protagonists of the lengthy story of such an apparently trivial matter as pitch darkness. I know of no other similar publication in this country or abroad that has dealt so graphically or succinctly with a question which, after four centuries of fumbling and chasing up blind alleys, was only solved in our lifetime. By being able to follow step by step this thorny path to understanding we acquire a much better grasp of the fundamentals of scientific research, and a basis to solving the other mysteries that developments in the natural sciences have lined up for us.

Prague, December 2010 *Jiří Grygar, Czech Academy of Science,*
Czech Astronomical Society,
Learned Society of the Czech Republic

Author's foreword

I have written this book for everybody who has not only been charmed by the majestic appearance of the starry night sky, but has also started to reflect more deeply on the structure and the physical nature of the universe. The text is formulated for "entry level", and is intended to be accessible for high school students and for all other non-experts without a deeper knowledge of astronomy. Nevertheless, I hope it will also provide something stimulating for readers whose interest in astronomy and in the history of science and philosophy is deeper.

I am obliged to astronomer Dr. Jiří Grygar (Czech Academy of Science, Institute of Physics) and to Prof. Petr Kulhánek (Czech Technical University in Prague) for their comments on the original Czech text. My special thanks belong to Gerald Turner for his translation into English.

Mysteriously self-evident

When the sun falls beneath the horizon, darkness fills the sky above our heads. Why? Why is the sky dark at night? If you don't know, don't worry. Scarcely anyone knows. And those that don't, probably don't know they don't know, because they've never asked themselves the question. "Why is it dark at night?" is a question that, for centuries, not even philosophers or natural scientists asked themselves, but nor did astronomers, who should have been best equipped to tackle it.

THE TRUE MYSTERY OF THE WORLD IS THE VISIBLE, NOT THE INVISIBLE.

Oscar Wilde

The night sky has fascinated the human race since time immemorial. Our attention is drawn by everything we can see in the sky, everything that glistens and moves there. What is that glow? What does it consist of? What makes it move and why? But most of the night sky is covered in empty darkness. It doesn't glisten and it doesn't move. It consists of nothing. It is not anything real, just absence of visual perception. Darkness forms only a monotonous backdrop, in front of which the cosmic theatre is played out. It was not until the Renaissance that astronomers realised that darkness is self-evident only in terms of our daily—or rather nightly—experience. In terms of the structure of the universe it is not self-evident at all. So it's not without interest. What was self-evident ceased to be so and became a mystery instead. A paradox came into existence. Astronomers started to look for an answer to the question "Why is it dark at night?" It was sought by Kepler, Digges, Galileo, Halley, Olbers and other astronomers. And not just astronomers: even the philosopher Friedrich Engels was interested in the question and the famous author Edgar Allan Poe tried to find an answer. The question of darkness at night continued to occupy the minds of cosmologists in the twentieth century. Is it still topical now in the twenty-first century?

Let's go back to the beginning and ask ourselves the question: Why isn't it obvious that it is dark at night? Why oughtn't it to be dark? And why is it dark nevertheless?

> *The mysteriousness of obvious phenomena eludes us. We are not surprised by them and don't investigate them. If they are to surprise us we have to render them "non-self-evident". It was a long time before human beings were interested in why things fell, and why they fell just in one direction. It only ceased to be self-evident when that famous apple fell on that distinguished head. That led to enlightenment: gravity was no longer self-evident and it became a matter for reflection and investigation.[1] The situation with magnetism had been quite the opposite; ever since ancient times people had been amazed by magnets. Thales of Miletus, known as "the first philosopher", was amazed by magnetism to such an extent that he abandoned his "naïve materialism" and offered gods and souls as an explanation of magnetism. And magnetism continues to fascinate even now. It became mysterious and magical because it was an unusual phenomenon. But like gravity, there was nothing curious about darkness—after all we encountered it even before we saw the light of the world.*

---

[1] The reader is no doubt aware that the head in question belonged to Isaac Newton, about whom his biographer William Stukeley wrote: "(Newton) told me, he was just in the same situation, as when formerly, the notion of gravitation came into his mind. 'Why should that apple always descend perpendicularly to the ground,' thought he to himself: occasion'd by the fall of an apple, as he sat in a contemplative mood: 'why should it not go sideways, or upwards? but constantly to the earth's centre? Assuredly, the reason is, that the earth draws it. There must be a drawing power in matter.'"

Billions of alien suns, or why it ought not to be dark at night

... IF THEY ARE SUNS HAVING THE SAME NATURE AS OUR SUN, WHY DO THESE SUNS NOT COLLECTIVELY OUTSHINE OUR SUN IN BRILLIANCE?

Johannes Kepler
(*Conversation with Galileo's Sidereal Messenger*)

If we look around us in the forest, what do we see? Trees upon trees. Whichever direction we look in, we see a tree. Trees fill the entire field of view around us (at ground level, at least). And the same should apply to the forest of stars in cosmic space: there should be one shining in every direction.[2] Stars should cover the entire vault of heaven with no dark spaces. And since Kepler was right when he assumed that the stars are alien suns like our own Sun with a capital S, their surfaces glow like the surface of the Sun. And so we should be surrounded by unbearable glare and heat. There should be no night and no day, just an inferno. Here we come to the crux of the matter: the fact is that at night there is neither light nor an inferno. It is cool and dark, and there are no fires of hell. (This refers to our properly heated central-European Hell. The Hell of the Nordic nations is an icy wasteland.) We are confronted with a contradiction with reality known as the dark night sky paradox. It is also called the photometric paradox, Olbers' Paradox, and sometimes Kepler's Paradox, Halley's Paradox, and surprisingly also the "light sky paradox".

---

[2] Otto von Guericke (1602-1686) was the first to compare the universe to a forest.

Peter Zamarovský

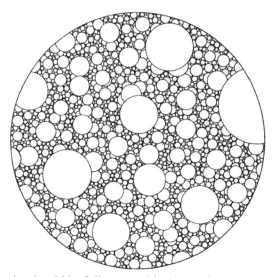

The whole sky should be fully covered by discs of stars (Harrison 1987)

Does something change when we move from a forest of trees to a forest of stars? Let us imagine ourselves surrounded by an infinite collection of concentric spherical shells, something like the layers of an onion, except that the structure is infinite. The shells are of the same thickness and they are so immense that each of them contains a large number of stars (so immense that the unevenness of their distribution averages out). Then the number of stars in each of the shells is commensurate to their volume and the volume in turn is commensurate to the area of the shell. So the number of stars increases with the square of the shell's radius, i.e. with a square of the distance from us. The intensity of the light from the individual stars, on the other hand, is inversely proportional to the square of their distance.[3] The two relationships disrupt each other so all the shells ought to contribute equally to the sky's radiance. However, the

---

[3] This laws of optics was formulated in 1604 by Johannes Kepler in his treatise *Supplement to Witelo, concerning optical astronomy* (*Ad Vitellionem paralipomena quibus astronomie pars optica traditur*), Frankfurt 1604. The law applies to a point source of light that is not absorbed by its surrounding, as well as to three-dimensional Euclidean space.

number of shells is infinite, so an infinite radiance should emanate from the sky . . .

But there is no infinite radiance emanating from the sky. Where was the mistake? What was wrong with our model? From our experience of a terrestrial forest it will probably occur to us that we mistakenly substituted geometrical points for the stars. Although it might not occur to as at first (or even second) glance, the stars are enormous spheres, which—unlike geometrical points—are capable of obscuring each other in the way tree trunks do. As a result we cannot see into infinity. We cannot see all the spheres or the infinite number of stars. So the sky's overall glow will not be infinite. But any rejoicing is premature: the fact that the stars can cover each other doesn't protect us from an infernal glow. The sky's glow might not be infinite but the sky would glow brighter than 90,000 Suns!

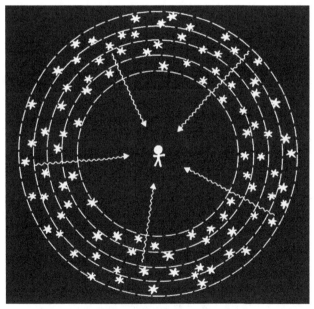

Concentric shells around observer (The shell model was introduced by Edmond Halley.)

> *"What if" speculations tend to be crudely simplistic, distorting and misleading. This equally applies to our analysis. The temperatures at the surface of stars reach thousands of degrees. That may seem a lot, but on stars that is not a high temperature. The temperature at a star's core is fifteen million degrees or more (according to the type of star). If the "inferno" spoken of earlier occurred the stars would not have anywhere to radiate their energy to. They would become overheated and the thermonuclear reactions that normally create heat only within their cores would spread throughout the entire volume of the stars and a mighty explosion would occur. Astrophysicists assume that such an explosion occurs when stars collide. (Unlike motor cars, stars rarely collide.)*

We have presented an example of how the universe would appear if it were infinite and uniformly (or at least randomly) filled with stars. If it really did look like that, we wouldn't be able to see it, because it would be too hot for us to exist.

> *A mere glance at the sky tells us that the stars are not uniformly distributed. The uniform arrangement of objects is only encountered where some factor causes the arrangement, such as intermolecular bonds in a crystal or a fruit grower's plan to plant trees in rows. If only random factors apply, a random, so-called "Poisson distribution"[4] ensues, in which dense and less dense areas occur in an irregular fashion. This happens when random factors prevail over those that promote an arrangement, such as when a crystal is heated up and the rapidly moving molecules collide, thereby disrupting the crystalline structure and causing the crystal to melt. (Examples of random distribution are the spatial distribution of trees in a forest, the number of deaths per day in London, the number of disintegrations of a radionuclide atom over a given period of time, etc. However, if we average out random distribution over a larger interval or area it comes close*

---

[4] That kind of distribution has been studied by Siméon Denis Poisson (1781-1840), French mathematician, geometer, and physicist.

Why is it dark at night?

> *to uniform distribution.) Maybe it is worth pointing out that, if, for instance, 3600 atoms of a given isotope decay per hour, this doesn't mean that one atom decays precisely every second. Therefore, for our purposes we can regard the random distribution of stars as uniform. In the case of the distribution of stars in space, as with many other natural elements, randomness is generally combined with a certain kind of regularity.*

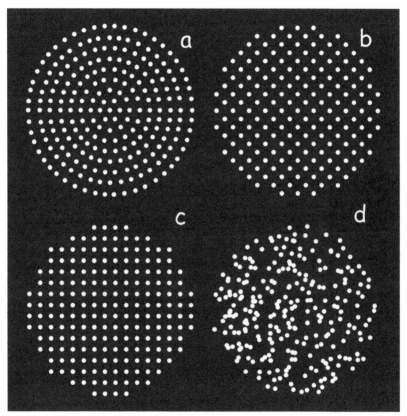

Some examples of uniform distribution (a, b, c) and random distribution (d)

So what do we see in the night sky? A vault covered in velvety darkness and on it myriads of stars?[5] That sounds poetic but it isn't

---

[5] A myriad is ten thousand. The ancient civilisations, such as the Greeks or Chinese, generally regarded ten thousand as a countless or infinite quantity.

the case[6]. With the naked eye we are able to count only two or three thousand stars (about six thousand in both hemispheres). No billions or myriads, no astronomical figure. We arrive at a bigger figure if we use a telescope, which enlarges the image of the sky. But enlargement of itself is not important—the reason we don't see the weak stars is not because they are small but because not enough light reaches us from them. To render stars visible it is therefore essential that the objective lens of the telescope should be capable of concentrating more light than the lens of the naked human eye. If the diameter of the objective is 5 cm, for instance, its area is some one hundred times greater than the area of our pupil and so it captures one hundred times more light.[7] Therefore with a telescope of this kind we can see stars that shine one hundred times more weakly. Had we the patience we would be able to count hundreds of thousands of them by now, in other words about a hundred times more than with the naked eye. Let's try it with an even bigger telescope, one with an objective measuring 50 cm in diameter (a powerful amateur telescope or a regular professional instrument). Such a telescope concentrates about ten thousand times more light than the eye. With it we can see about ten thousand times more stars—about twenty million of them. We would never finish counting them in our lifetime.

The number of visible stars increases more or less in proportion to the area of the objective. The weaker the stars are the more there are of them in the sky. This might support the assumption that the stars are distributed uniformly: an objective ten times bigger concentrates a hundred times more light, which means that stars that are one hundred times weaker are now visible, i.e. stars from the shell ten times further away. And there are roughly one hundred times more of those stars, because the volume of the shell ten times further away is hundred times bigger and therefore contains one hundred times more stars. However, as we shall soon see, this relation ceases

---

[6] And due to increasing light pollution the velvety darkness doesn't even apply either.

[7] The diameter of the pupil—i.e. the "objective"—in the human eye is 2-8 mm. Let us say 5 mm for simplicity's sake. Since the diameter of the objective is ten times greater, its area is one hundred times greater.

Why is it dark at night?

to apply in the case of objectives with very large diameters—in other words, those needed for observing more distant parts of the universe.[8] So with a telescope we can see a greater number of stars overall. Not all at once, however, because as the image is magnified, so the field of view is narrowed. And we continue to see more or less the same picture—the stars as shining points of light with darkness in between them. The existence of darkness between the stars is not dependent on the size of the telescope or its power of magnification.

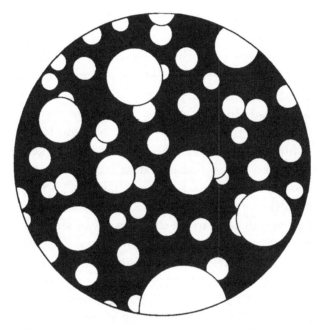

In reality, there remains dark sky between the stars (Harrison 1987)

I expect that everyone who looked through an astronomical telescope for the first time was disappointed to discover that the stars look like points of light even through a telescope. (If we exaggerate the telescope's magnification, the stars become shining circles surrounded by rings and with a coloured fringe and sometimes something like dancing amoeba. These mis-shapes

---

[8] The greatest density of stars in the field of view is obtained by using a low-power "rich field telescope" with a diameter of 10-20 cm.

have nothing to do with the actual shape of the stars, of course. They are caused by the optics of the instrument and by atmospheric turbulence.) And it would surely have disappointed even Tycho Brahe had he lived a few years longer and had a telescope available to him. He had estimated that the apparent diameter of the stars was about two minutes of arc. (A minute of arc (MOA) is a unit of angular measurement. Two minutes of arc would be the size of a pea at a distance of about eleven metres. To see it at an angle of one second of arc, we would have to view it from a distance of 1.4 kilometres.) If this were so we would only have to enlarge it fifteen times in order to see it as big as the Moon's disc. The apparent diameters of all the stars largely lay below the resolving power of telescopes located on Earth. (The situation is better with the planets. With the telescope we can observe their small discs or sickles, and if we are lucky we can even see details on their surface. Venus has the largest apparent diameter, which can be over one arc minute.) Take the case of Sirius, for instance, which is the clearest star in the whole sky. This near and also giant star appears to us to have an apparent diameter of 6 thousandths of an arc second. (For a pea to appear to us like Sirius it would need to be 200 km away and have the same temperature as Sirius, i.e. 10,000 °C.) In practice, the resolution limit of astronomical telescopes is usually about one arc second and in the best cases several tenths of a second. If we wanted to see at least some of the stars as discs, we would need a thousand times more powerful telescopes. But even they would not help us much. We live at the bottom of an ocean of air, where masses of air mingle at various temperatures and therefore also at various densities (and optical refractive indices). Therefore the resolution of big telescopes (starting from 10-20 cm in diameter) tends not to be limited by the performance of the optics but precisely because of blurring due to atmospheric turbulence. So astronomers cannot see the stars as discs even with big telescopes. All they can do is measure the intensity and spectral composition of their light.

The situation could be solved by telescope located outside the atmosphere as witness the fine photographs from the Hubble Space Telescope. For that unique instrument the nearest stars were no longer points of light. In 1995 this telescope managed to photograph the disc of the star Betelgeuse, one of the two brightest stars in the constellation of Orion. This supergiant (not a metaphor, but its official designation) is about 800 times bigger than the Sun and is at a distance of 130 light years from us. Its apparent diameter is 90 thousandths of an arc second, i.e. 1/500 of the apparent diameter of Jupiter or 1/2000 of the Moon's disk. However, rapid strides have been made in recent years also in the development of large earth-located telescopes. To a certain extent, the new equipment, i.e. so called active optics, is capable of compensating for the shimmer and defocusing of the image caused by the non-homogeneity of Earth's atmosphere. So far this technology mostly concerns only a small number of very large instruments.

Why it is dark at night

The darkness of the night sky provides us with important testimony about the structure of the universe: Whatever the universe is, the way it is ordered in space and time ensures that there is darkness at night. In the first place, this implies that the universe cannot be infinite and at the same time filled with eternally shining stars in a regular (even random) arrangement. It must be something different. But what?

IF THERE WERE NO SUN, ON ACCOUNT OF THE OTHER STARS IT WOULD BE NIGHT.

Heraclitus B99[9]

Cosmos, Myth and Logos

How is the universe structured? How and when did it come into existence? How does it relate to us, human beings? Curious homo sapiens has most likely been ruminating on such questions ever since the stone age. And what humans didn't know they made up for with flights of imagination, giving rise to myths, which for thousands of years governed not only the opinions but also the daily lives of generations of our ancestors. The terrestrial and superterrestrial worlds were both viewed from a mythical perspective.

A heavenly vault filled with stars revolved above people's heads and among the fixed stars there moved the wandering stars—the planets, which included the Sun and the Moon. (Astrologers still include them.) Since the planets moved they were considered to

---

[9] "B" denotes an authentic fragment, "99" the number of the fragment according to the standard Diels-Kranz numbering system.

be animate and soulful beings, both ethereal and divine. Thanks to the fabulous imagination of our ancestors those ancient gods have survived right down to the present day: towards evening we can admire glittering Venus, the goddess of love (known in Greek as Aphrodite). Sometimes we can catch sight of Mercury in the close vicinity of the Sun. Reddish Mars, the god of war boasts two sons, Phobos (signifying panic or fear) and Deimos (dread) which orbit it as moons.[10] The supreme god Jupiter (Zeus in Greek) is perhaps the most beautiful of the planets, while no less remarkable is Saturn, with its rings. The planets discovered in modern times were named in accordance with the classical tradition: Uranus, Neptune and Pluto. After their deaths even mythical characters took up residence in the sky: Hercules, Cassiopeia, Andromeda, Castor and Pollux, Orion . . . They look down on us in the form of vast constellations. The sky itself received as its ruler the cruel god Uranus. And Urania then became the muse of astronomy, because before it became subject to the exact sciences astronomy used to be an art and therefore had its muse. For their part, the ancient Egyptians entrusted dominion over the heavens to the goddess Nut. And Ra, the sun god, and the Sun itself, was her son.[11] In ancient India, Varuna was the goddess of the heavens . . .

The mythical image of the world became a source of inspiration for thousands of artists and it has still lost none of its allure. Myths are not only entertaining, they also contain edifying tales to do with the lives of individuals and societies. We can find within them archetypal characters and situations that have shaped people's consciousness and unconscious. But it would be pointless to seek within them any critical scientific approach. A realisation of the limitations and naivety of those myths already started to emerge in the ancient world and humanity started to outstrip mythology. It invented critical thinking and was surprised at its enormous potential. Soon philosophy was

---

[10] They were added to the ancient pantheon later. The moons of Mars were discovered in 1877.
[11] Just to complicate matters, the function of sun god was occasionally assumed by other gods, such as Atum.

born and the various branches of science gradually developed from it. *Sophia*—wisdom—ceased to be knowledge of myths and became the possession of reason—*logos*. Humanity was coming of age.

> *The Greek word logos means "word" or "meaningful speech", but also "reason" and even "mathematical fraction", i.e. a rational number. The renowned philosopher Heraclitus of Ephesus used the term logos to describe universal order. The concept was later developed by the Stoics, who regarded logos as a universal creative principle, a generative force, and even a god. And logos even entered Christian tradition. The Gospel of St John begins with the words "In the beginning was the word"—"logos" in the original Greek.[12] In Christian understanding logos thus becomes the Son of God. He was the means by which God created the world and through him God is revealed to the world. And so logos returned to the domain of myth . . .*

The Universes of the Earliest Astronomers

So let us turn to the models of the universe based on a rational approach, observation and *logos*. Let us see how well they coped with our paradox, i.e. whether and how they explain nocturnal darkness.

The earliest philosophers started to speculate about the universe. Since the main subject of their interest was *physis* they have been termed "physicists". The word *physis (natura* in Latin) signified nature (particularly living, vegetable nature), naturalness, essential qualities, innate disposition. Only a fraction of the teachings of those "physicists" has come down to us—in quotations by later authors. This allows us to reconstruct only a tentative and partial picture of their teachings.

---

[12] A thought for the philosophically minded: The Chinese translate *logos* in the Gospel by the word *tao*.

The first to make his name in the field of astronomy was Thales of Miletus. Thanks to his successful prediction of a solar eclipse he achieved the fame and esteem accorded to prophets. He achieved that success thanks partly to the Chaldean astronomical tables and partly to a fair amount of luck. (The tables predicted only the possibility of an eclipse.) Thales believed Earth was an island floating in an ocean around which the heavenly spheres revolved. How the spheres managed this when the ocean was infinite is unclear.

The view of Thales' younger colleague Anaximander was that Earth floated freely in space. It retained its balance because the "impulses" from various sides cancelled each other out because of symmetry.

> *If that interpretation is correct, then Anaximander had a premonition of the law of universal gravitation two thousand years before Newton—at least in a qualitative sense. Let us recall that an entire millennium after Anaximander, St Augustine regarded as fools those who thought the other side of our planet was inhabited. They failed to understand that the antipodeans would have to walk upside down! Similar ideas were even asserted by the famous astronomer Claudius Ptolemy. If human knowledge hadn't wasted so much time in blind alleys we might have been much further ahead . . . The question is where?*

According to Anaximander the individual heavenly bodies are located at various distances from Earth. So the visible universe is not the surface of some celestial dome, but has depth—it is "3D". There were other early astronomers who toyed with the idea of cosmic space extending into the depths, but it took two thousand years for this to take root. The celestial spheres were finally done away with by the Renaissance, which opened up the depths of the cosmos and assigned the heavens to the theologians for good.

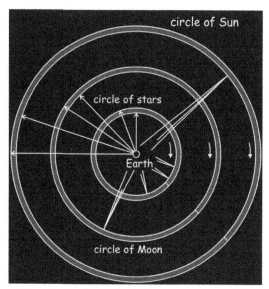

The cosmos of Anaximander had depth, various celestial objects were at various distances

We must excuse Anaximander's failure to guess correctly the order of the cosmic bodies: he placed the sparkling stars nearest, then the cold Moon, and the fiery Sun was supposed to move in the distance. Maybe he placed the stars nearest because he believed them to be sparks that had been given off when Earth was being formed. It's not surprising: stars really do look more like sparks than huge balls of fire.

These notions were the inspiration for the legendary Pythagoras (c. 572-494 BCE) along with his pupils and followers. It was already obvious to the Pythagorean astronomers that Earth was neither a pancake nor an island floating in an ocean, but was an enormous ball floating freely in cosmic space. Some of them even rose above egocentric geocentrism and realised that our native planet wasn't so exceptional and was just one of many similar cosmic bodies.

> At the end of the 5th century BCE the Pythagorean Philolaus of Croton described a model of the universe in which Earth was not at the centre. The planets, including our own, revolve around a central fire—a "cosmic hearth". (The central fire lies beneath our feet and is covered by the globe.) And in order to demonstrate that he had not forgotten about the seat of the gods, he placed the dwelling of the supreme deity Zeus down there. (Two thousand years later, Thomas Digges also had a problem with where to place the seat of God—but we are rather jumping ahead.) Nowadays it is obvious to us that Philolaus' abandonment of geocentrism was a step in the right direction. But at the time it was a premature birth.

Like Anaximander, the Pythagoreans located the cosmic bodies at various distances from Earth. These distances created a harmonious order (similar to musical tones) and so perfect harmony and order reigned in the entire Pythagorean universe.[13] Pythagoras (though other accounts claim it was Philolaus) named the entirety of everything "the cosmos". The word itself was older and is to be found, for instance, in Homer's Iliad. It meant order, an organised formation, discipline, tidiness, ornament. It is also the root of the word "cosmetics"—beauty products intended to impart order, and hence beauty, to the human exterior. The Pythagoreans called the universe "the cosmos" because they presumed that within it there was order and harmony. From "cosmos" we also have "cosmology" (and also "cosmography")—the study of the structure of the universe, and "cosmogony"—the study of the origins and evolution of the universe. Since the origins and evolution of the universe are intimately bound up with its structure, cosmogony was subsequently subsumed into cosmology.

Emphasis on the existence of mathematically describable order—the cosmos—is Pythagoreans' greatest contribution to nascent philosophy

---

[13] The Pythagorean concept of universal order had its roots perhaps in ancient Egypt. It subsequently had a major influence on Johannes Kepler.

and future science. Philosophy is not to do with particulars or specifics but with order. Without the presumption of order, science would not be possible, not even astronomy. Nor would there be astrology, for that matter. Whatever we think of it, astrology definitely presumes some kind of universal order. It's a pity that so little information about the Pythagoreans' concept of the universe as a whole has been preserved. We do know, however, that they endeavoured to understand everything on the basis of a finite number of specific basic elements. They feared infinity. In spite of that "apeirophobia"[14] however, it would seem to have been Pythagoras' follower Archytas of Tarentum (c. 400-365 BCE), who was the first to formulate clearly that the universe ought to be infinite. He illustrated the snags implicit with a "cosmic frontier" by using a "thought experiment". He asked the question:

"IF I HURL A JAVELIN ACROSS THE FRONTIER OF THE UNIVERSE, WHAT WILL HAPPEN TO IT? WILL IT BOUNCE BACK? OR WILL IT DISAPPEAR FROM THE WORLD?"[15]

The ancient atomists, such as Leucippus (c. 500-440 BCE) and Democritus (c. 460-370 BCE), also speculated about the global structure of the universe. Their universe also had no limits, no "final sphere", no heaven. It was infinite, or at least boundless.[16] According to Leucippus it consisted of "fullness and emptiness". Emptiness (a vacuum—*kenon* in Greek) created space; fullness was represented by the individual "worlds"—*cosmoses* in Greek (consisting of smaller "fullnesses", this order ended with the elementary "fullnesses"—atoms). We must be circumspect when interpreting the words of the atomists; for them *cosmos* did not represent the universe as a whole, but only one "ordered part" of it.

---

[14] "Apeirophobia" signifies fear of infinity, from *apeiron*—infinity, and *phobia*—fear

[15] From Simplicius (6th century CE) commentary to Aristotle *Physics*.

[16] At the time there was no distinction between these concepts.

Why is it dark at night?

> ... THERE ARE INNUMERABLE WORLDS (COSMOSES), WHICH DIFFER IN SIZE. IN SOME WORLDS THERE IS NO SUN AND MOON, IN OTHERS THEY ARE LARGER THAN IN OUR WORLD, AND IN OTHERS MORE NUMEROUS ... THE INTERVALS BETWEEN THE WORLDS ARE UNEQUAL; IN SOME PARTS THERE ARE MORE WORLDS, IN OTHERS FEWER; SOME ARE INCREASING, SOME AT THEIR HEIGHT, SOME DECREASING; IN SOME PARTS THEY ARE ARISING, IN OTHERS FAILING. THEY ARE DESTROYED BY COLLISION ONE WITH ANOTHER. THERE ARE SOME WORLDS DEVOID OF LIVING CREATURES OR PLANTS OR ANY MOISTURE.
>
> *A40 from Hippolytus*[17]

What are we to understand by these "worlds"—cosmoses—now? It is possible they could be planetary systems or planets, most likely extrasolar, which orbit other suns. Of course it would also make sense to interpret "worlds" to mean "entire universes". We would then end up with a concept of a multiverse, a universe of universes. These concepts are also being developed as part of present day cosmology.

> *A historian of philosophy would probably have reservations about such modern interpretations. It is anachronistic to attribute present-day concepts to the ancient sages, and therefore a grave transgression against the principles of historiography. Nevertheless I find such anachronistic interpretations the most appropriate.*

---

[17] The principal work of Hippolytus of Rome (170-235) was the *Refutation of All Heresies (Refutatio omnium haeresum)*. It is a compendious polemical work, which catalogues both pagan beliefs and 33 heretical Christian systems.

> *The German philosopher Georg Friedrich Hegel (1770-1831) put forward the view that the history of philosophy and philosophy are one and the same thing, an idea that continues to have a baleful impact on the teaching of philosophy. It is my opinion that we should not restrict ourselves to the historical context, to "what they actually meant at the time". What is appealing about philosophical texts is precisely the fact that they permit or even invite many different interpretations. Therefore we should not be afraid of "superinterpretation". We should ask ourselves what a particular philosopher says to us now. Many philosophers deliberately express themselves ambiguously. Their use of allegory or even obscure language irks their more exactly-minded colleagues who demand clarity in all things. But it gives readers scope to decipher the meaning of the text and project onto it their own views, impose their own meanings, their own interpretations, their own "reading". Ambiguities can thus be a fertile or even fecund element in the text. This typically applies to mythological and religious texts, and even to some formulations of physical theories.[18] Stephen Hawking and Jacob Bekenstein have demonstrated, for instance, that the formalism which was developed in the nineteenth century to describe steam engines and heat phenomena (i.e. thermodynamics) can be adapted to describe black holes as well. In the process, the original thermodynamic quantities (heat, entropy, etc.) acquire an entirely new—"anachronistic"—significance.*

The atomists' revolutionary contribution was the idea of the development of the universe, or at least the development of individual "cosmoses". Cosmoses come into being, evolve and die. Democritus' words sound as if they come from a current textbook on astrophysics.

---

[18] The branch of knowledge dealing with the interpretation of texts is known as hermeneutics, and it was originally concerned with religious texts.

The atomists came up with a number of ideas that are still pertinent, but they were rather unlucky: shortly afterwards their theories about physics fell by the wayside and they were not particularly appreciated in later times, being overshadowed by such philosophical giants as Socrates, Plato and Aristotle. (We will come back to their ideas about the universe later.) It was Plato who started to put the lid on atomism by systematically ignoring the atomists in his philosophical works. According to Diogenes Laertius, he actually intended to gather together their writings and burn them. But he came up with the idea too late—thankfully. Although Aristotle seems to have referred fairly objectively to the atomists, he asserted a quite different approach.

THE NOTION OF THEIR INDEFINITENESS (NUMBER OF WORLDS) IS CHARACTERISTIC OF A SADLY INDEFINITE AND IGNORANT MIND.

Plato, *Timaius and Critias*, IV, 55d
(taunt against atomists' opinions)

The Universe of the Epicureans

Nevertheless, the atomists' teachings managed to survive in the shadow of the heavyweights. They were passed down from teacher to pupil and, thanks to a certain Nausiphanes, they came to the attention of a lively-minded young man by the name of Epicurus (341-270 BCE) and they fired his enthusiasm. Unlike other philosophers, Epicurus had realised that philosophy needed a firm basis, and that basis was physics. He correctly surmised that the atomists' approach to physics had been the most well-thought-through and thoroughgoing, and so he breathed fresh life into dormant atomism, developing it and fitting it into the broader context of his philosophy.

EACH OF THESE PHENOMENA (ASTRONOMICAL PHENOMENA) ALLOWS SEVERAL DIFFERING EXPLANATIONS FOR ITS CREATION AND ITS NATURE, ALL OF WHICH MAY AGREE WITH VISIBLE EVIDENCE.

Epicurus, *Letter to Pythocles 86, 87*[19]

Like the early atomists, Epicurus regarded the universe as an unbounded void more or less evenly filled with stars. He argued in a similar fashion to Archytas: If the universe had limits, what would be beyond those limits, in addition to it? Hence Epicurus' universe is boundless. It also has no centre, or any other important place that might be intended for us people.

. . . THE UNIVERSE IS INFINITE. FOR THAT WHICH IS FINITE HAS AN OUTMOST EDGE, AND AN OUTMOST EDGE CAN ONLY BE FOUND IN COMPARISON TO SOMETHING BEYOND IT. BUT THE UNIVERSE CANNOT BE SO COMPARED, HENCE, SINCE IT HAS NO OUTMOST EDGE, IT HAS NO LIMIT; AND SINCE IT HAS NO LIMIT, IT MUST BE UNLIMITED AND INFINITE.

Epicurus, *Letter to Herodotus (DL, X, 42)*

Epicurus realised that there existed no knowledge of the origin or eventual demise of the world and gods could not be relied on, not only in the world of humans but also in cosmogony. So he opted for the simplest solution—that there had been no beginning of the universe and that there would be no end.

---

[19] from Diogenes Laertius (DL)

... THE UNIVERSE HAS ALWAYS BEEN AS IT IS NOW, AND ALWAYS WILL BE, SINCE THERE IS NOTHING IT CAN CHANGE INTO.

Epicurus, *Letter to Herodotus* (DL, X, 39)

Epicurus' cosmological concepts have survived until today even though his authorship is slightly forgotten. Nowadays we call them "cosmological principles". A cosmological principle expresses the spatial homogeneity of the universe. Moreover, the so-called Perfect Cosmological Principle asserts temporal homogeneity, in other words, that "essentially everything has always been and will always be the same", precisely as Epicurus stated.

> *The cosmological principle was rediscovered in 1948 by the British physicist E. A. Milne (1896-1950). It is valid on an appropriately large scale, of course. However, the Perfect Cosmological Principle, which demands also the homogeneity of the universe in time, seems not to be valid. (See later.)*

The chief advantage of the Epicurean model of the universe would seem to be its rational level-headedness. It eschews speculation about the creation or the demise of the world, it requires no heavens; an outlying sphere of fixed stars or other unreal assumptions. It needs no mythological explanations. The universe had simply always been and always would be.

JUST AVOID THE MYTHICAL EXPLANATIONS, WHICH YOU WILL DO AS LONG AS YOU FOLLOW THE EVIDENCE OF THE SENSES TO GET INDICATIONS ABOUT WHAT IS UNSEEN.

Epicurus, *Letter to Pythoclus* (DL X, 104)

But neither Epicurus, nor his pupils and followers seem to have noticed that their concept of the universe has a serious flaw. It is this model that harbours the paradox that we are dealing with in this book: If the universe really was like that there would be no night, the sky would glow and things would be really hot down here.

The Universe of the Stoics

Operating alongside the Epicureans was the school of the Stoics. In a sense they were at the opposite pole to the Epicureans. They asserted not only austere rationalism, but also an ascetic lifestyle free of emotion and passion. Like the Epicureans, the Stoic scholars were interested in the physical world. While their physical knowledge might appear meagre from a modern standpoint, their fundamental approach was far-sighted. Their main contention was that nature and human beings are governed by a common order—the *logos*.

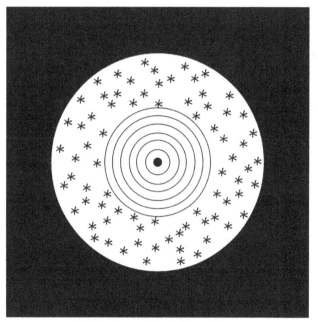

The Stoic model of the universe

The Stoics' ideas also related to the expanses of the universe. Although they regarded cosmic space as infinite, the stars were not scattered everywhere. They were just an island within otherwise empty space. They placed Earth—spherical of course by then—in the middle of that starry island.[20] Unlike the Epicureans' universe, the Stoics' universe wasn't eternal. It emerged from fire and was destined to return to a fiery state in time. The entire cosmic cycle—*ekpyrosis*—repeated itself over and over again. Noteworthy from our viewpoint is that in the Stoics' universe no dark night sky paradox arises—assuming that there are few enough stars and that they are sufficiently distant from each other. (Remember that in the Epicurean universe there was an infinite number of stars. In that case the paradox applies however distant the stars are from each other.) The black sky that we see between the stars is the empty space lying beyond the starry island. Were the Stoics aware, perhaps, of the dark night sky paradox? Did they deliberately create a cosmic model that avoided it? By now that is simply a matter of speculation.

> *Let us return to the concept of the "cosmic cycle". A cycle is a process that is repeated, a circular event, such as the life of animals or people. Does our universe repeat itself? Is it cyclical, and if it is, does it repeat itself exactly or only approximately? The idea of a repeating cosmic cycle was an elegant solution to the question of what preceded it and what will follow it. A developing cosmos is transformed into an everlasting cosmos that never came into being and would not come to an end, but instead repeats itself over and over again.*
>
> *Let us look at the concept of the "cosmic cycle" in greater detail. If the state of the entire universe is to repeat itself exactly, two conditions must be fulfilled:*
>
> *1.   The repeated state must be precisely the same (otherwise it would be no recurrence)*

---

[20] Archytas of Tarentum had most likely already asserted a similar cosmic model.

2. There must be something that distinguished the initial state from the recurrent state. Only then can it be asserted that the repeated state is not the same as the initial one. (According to Leibniz's principle of the identity of indiscernibles, two entities that have precisely the same properties are identical.)

But these two conditions contradict each other. Since the recurrence ought to involve absolutely everything, nothing different can be found and in principle the repeated state cannot differ from the initial one. Nothing exists that "remembers" the past cycles. The "interruption of memory" is due to the ekpyrotic fire that "resets" all memories. Instead of a cyclical (or oscillating) universe it would therefore be more appropriate to talk about a "temporally closed universe" or about "closed time".

Going back two thousand years to Ancient Rome, we find the Epicurean Lucretius Carus writing about a recurrent universe (i.e. the "rival" Stoic view):

FOR SHOULDST THOU GAZE BACKWARDS ACROSS ALL YESTERDAYS OF TIME THE IMMEASURABLE, THINKING HOW MANIFOLD THE MOTIONS OF MATTER ARE, THEN COULDST THOU WELL CREDIT THIS TOO: OFTEN THESE VERY SEEDS (FROM WHICH WE ARE TO-DAY) OF OLD WERE SET IN THE SAME ORDER AS THEY ARE TO-DAY-YET THIS WE CAN'T TO CONSCIOUSNESS RECALL THROUGH THE REMEMBERING MIND. FOR THERE HATH BEEN AN INTERPOSED PAUSE OF LIFE, AND WIDE HAVE ALL THE MOTIONS WANDERED EVERYWHERE FROM THESE OUR SENSES.

Titus Lucretius Carus, *De Rerum Natura*, III, 849-858

> With the passing of Antiquity the cyclical models of the universe were forgotten. Christianity stressed the opposite view, namely a linear development of the world, "linear time": the world was created, it is currently in progress, and sooner or later it will come to an end. Christ will return, there will be a Day of Judgement and eternity. Cyclical models were not revived until the twentieth century. [21]

In later ages the models of the Epicureans and Stoics were marginalised, like the teachings of the old Atomists. Europe did not recall them until the Renaissance period after the final demise of mediaeval Aristotelianism. They then underwent several reincarnations, modifications and subsequent decline.

---

[21] However, in Hinduism the idea of a recurrent universe still survives. Many philosophers, including Friedrich Nietzsche, also speculated about an eternal return.

The Aristotelian-Ptolemaic Cosmos

We have mentioned the most interesting cosmic models of ancient times. In various aspects the concepts of the Pythagoreans and Atomists, and later, the Stoics and Epicureans were similar to the present concept of the universe. There was a major gap in our historical extemporisation, however. This concerns the dominant astronomical concepts in the Classical period of Greek history which had a major influence on the spirit of Europe as a whole in the Middle Ages. From the point of view of cosmology, out of the three giants of philosophy—Socrates, Plato and Aristotle—only the last one was significant de facto.

> *Socrates, the first of the illustrious trio of philosophers, ignored "stellar mysteries" on principle. He regarded them as fatuous speculation with no practical purpose and incapable of being solved. This argument led to the rejections of cosmology as late as the mid 20th century! (There was also some rational reason for this as we shall see later.) Socrates contribution to cosmology was only indirect, namely his consistent assertion of rationalism and a method of reasoning based on mutual discussion.*
>
> *Even Plato, the second of the trio, did not tackle stellar mysteries, with one exception, his dialogue Timaeus. In it he paints a sort of mythical-allegorical picture, in which the cosmos is interpreted as a kind of animal. This was a fantasy of Plato like his story of Atlantis, which is told in the related dialogue Critias. It so happened that this was the only writing of Plato's known to mediaeval Europe, and many took it to be a seriously intended ancient cosmology. (The same fate was suffered by Plato's story about Atlantis. That is still taken seriously by some enthusiasts even nowadays.)*

So let us deal straight away with Aristotle and take a look at how the greatest scholar of ancient times conceived the universe. He didn't build his model on a greenfield site, but on foundations already laid by generations of earlier mathematicians/astronomers. Aristotle's model survived him and was improved on. The main credit for that goes to the astronomers from the famous Museum of Alexandria in Egypt. The most important of them were Hipparchus (2nd century BCE) and Claudius Ptolemy (2nd century CE), who supplemented the model by the addition of so-called epicycles.

According to Aristotle our round Earth is set at the centre of the Universe. Around it revolves a complex system of concentric crystalline spheres, on which the planets, including the Sun and Moon, are located. It was no easy task to describe the complex movements of the planets with the aid of celestial spheres. Astronomers therefore introduced several spheres that rotated around non-parallel axes in such a way that the axes of the inner spheres were fixed in the outer sphere. In this way a planet's motion consisted of several rotational movements. At the time of Aristotle it was thought that there were up to 55 spheres. The Earth and heavens were uncreated, unbegotten and eternal.

Before Aristotle these cosmic spheres were probably regarded simply as geometrical aids to describe the movement of the planets. It would seem that Aristotle was perhaps the first to regard them in material terms, as "crystalline" spheres (i.e. composed of a transparent substance). Amazingly, the concept of material cosmic spheres survived two whole millennia. The first to express doubts about it was the polymath Pietro d'Abano (c. 1257-1315), who expressed the view that cosmic bodies moved freely in space. It was not until the 16th century that Tycho Brahe finally dealt the theory its death blow. He demonstrated that comets flew through the celestial spheres without colliding with them. However, the transparent spheres have no influence on the darkness of the night sky. What is important is what lies behind them. According to Aristotelian notions, there is an overarching final sphere, the sphere of the fixed stars, which delimits and encloses our universe. And beyond it there is absolutely nothing. According to other modifications the stars are simply holes in the last sphere which admit the light of the fire blazing beyond the sphere. The universe itself is a kind of bubble in an enormous celestial fire. Thus the imagination of astronomers was once more enthralled by myth . . .

But let us return to the question: What is beyond the last celestial sphere? Is there really nothing? In other words, is the sphere a frontier, the end of the universe? But what does the "frontier of the universe" mean? This notion was already challenged by Archytas of Tarentum. The sphere of the fixed stars is thus evidently a weak point of the Aristotelian cosmos. (A physicist of today would probably call it a geometrical singularity.) And yet it had remarkable staying power. It survived the fall of geocentrism and even the crystalline spheres of the planets. It discreetly transferred itself to Copernicus' model and even Kepler accepted it in a modified form.

In late Antiquity and the Middle Ages the Aristotelian-Ptolemaic model almost achieved official status. This was assisted not only by the renown of Aristotle and Ptolemy but also by the mathematical excellence with which the model was elaborated. It provided very precise predictions of the planets' positions. No doubt its geocentrism was a psychological advantage, deriving as it did from people's innate egocentricity, and it also suited religious dogma that stressed the central position of Man as the image of God. The most important change wrought upon the Aristotelian-Ptolemaic model by mediaeval scholasticism was the ending of cosmic timelessness. The universe was created and would come to an end at a specified time. Christian dogma thus introduced two further "singularities": beginning and end, describing them as supernatural, divine acts.

> *Just as in all human notions, so also notions about the way the cosmos is organised reflect not only the object—the universe, but also the subject itself—human beings. We each see ourselves as the centre of all action. Innate egocentricity is also reflected in the fact that everyone regards his native sod, his country, as the navel of the world (e.g. ancient China as the "Middle Kingdom"). Later "our" planet became the centre, then "our" Sun, then "our" galaxy. However, to our chagrin, the centre kept moving away from us until it disappeared altogether. And with it our significance seemed to be waning . . .*

But let us return to our topic. How does the fact of the dark night sky fare in the Aristotelian-Ptolemaic model? The solution it suggests is suspiciously simple: The dark sky between the stars—that's the last sphere. The sphere is painted black and there aren't enough stars on it. So it is dark at night.

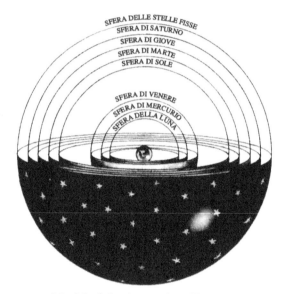

Model of the Aristotelian Universe

The Renaissance and the Return of the Epicurean Universe

At the beginning of the 15th century Aristotelian philosophy ran out of steam. Europe started to realise that ancient wisdom was not only Aristotle, his disciples and his commentators. The Renaissance was on its way. Platonism, Atomism, Epicureanism and Stoicism all saw a revival. In 1417 Cardinal Poggio Bracciolini discovered a remarkable manuscript—Lucretius' poem *On the Nature of Things* (*De Rerum Natura*). Europe learned from it many new features of Epicurus' ancient teachings. And the Epicurean model of the cosmos reappears on the scene, in other words, the notion of unending space filled with stars. In 1440 Cosimo de Medici founded the Academy in Florence. The institution sought its inspiration in Plato's school of the same name. That same year Cardinal Nicholas of Cusa published a work with the paradoxical title *On Learned Ignorance* (*De docta ignorantia*). In it, Nicholas takes issue with the Aristotelian concept of the celestial spheres and argues against a finite universe.

> *Nicholas of Cusa did not maintain that the universe was infinite (for him, as a theologian, only God was infinite), but simply that it was "indeterminatum", i.e. unlimited, without borders and without a centre. It might therefore seem that his conclusion is in line with several modern cosmologies working with closed non-Euclidian geometry: whilst the universe is finite it is without borders. Nicholas of Cusa's assertion was, however, framed more in epistemological terms, and its meaning was that the universe is undefined, i.e. it cannot be perceived rationally. That would comply with the paradoxical expression "learned ignorance", which clearly refers to Socrates' "I know that I know nothing".*

According to Nicholas of Cusa, the Earth has a circular orbit. It is unclear, however, whether he meant a circular trajectory around the Sun. The stars were (in accordance with Democritus and Epicurus) distant suns similar to our own. Nicholas' views sound revolutionary and progressive but they provoked no storm in their own day. His contemporaries did not understand them and therefore ignored them. They were not to receive a response until after Cusa's death. Both René Descartes and Johannes Kepler referred to them. And when tried by the Inquisition, Giordano Bruno defended his heretical cosmological view precisely by citing Cardinal Nicholas of Cusa as an authority.

In 1475 a Latin translation of Diogenes Laertius' biographical work *The Lives and Opinions of Eminent Philosophers* (*Laertii Diogenis Vitae et sententiae eorum qui in philosophia probati fuerunt*), in which the author relates the lives and deaths, but above all the ideas of the best-known Greek philosophers. It continues to be a valuable source of information about the philosophy of Antiquity.
It was not until Copernicus that cosmology was truly shaken to its foundations. The crucial moment was when Copernicus stood Ptolemy's objection to our planet's rotation on its head. Ptolemy had maintained that if the Earth were to move in a daily rotation, things would be thrown off from its surface by (centrifugal) force. But Copernicus realised that if the whole celestial sphere rotated it would be subject to an even greater centrifugal force, since its radius was much bigger than that of the Earth. Therefore the things contained are more likely to rotate than that which contains them, the part rather than the whole, our little planet rather than the enormous celestial sphere.

AFTER ALL, MUCH MORE WE WOULD WONDER, IF THERE IN A PERIOD OF TWENTY-FOUR HOURS SHOULD ROTATE SUCH HUGE VASTNESS OF THE WORLD RATHER THAN ITS TINY COMPONENTS, SUCH AS OUR EARTH.

Nicolaus Copernicus

An even more revolutionary change was heliocentrism—shifting the centre of the universe to the Sun.[22] Copernicus' simple yet brilliant reversal simplified the entire view of the universe. It made it easier to describe the movements of the planets and subsequently enable their physical explanation.

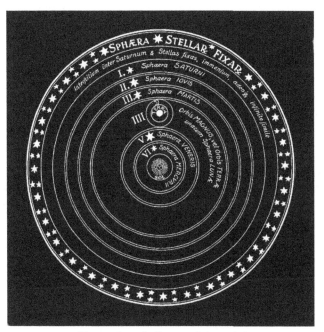

Copernican heliocentric system

Copernicus halted the rotation of the sphere of fixed stars, but that was as far as he went. He did not abolish the sphere of the fixed stars i.e. the borders of the universe. He did not scatter the stars into infinite space. Nonetheless Copernicus' "shift of the centre" was more of a breakthrough than its author himself realised. It evoked a return to the old questions of the Epicureans: If the Earth is not the centre of the universe why should any other place be the centre? And does the universe even have a centre?

---

[22] Copernicus did not create a "precisely" heliocentric system. He placed the centre of the Earth's circular orbit a short way away from the Sun, seeking, by means of that asymmetry, to justify the irregularity of the Sun's yearly movement across the sky.

Paradoxes around the Dark Night Sky Paradox

The Renaissance thus abandoned the out-dated Aristotelian-Ptolemaic universe and returned to the Epicurean universe. It seemed more sensible: the creation of the universe, its extinction and its boundaries presented no problems to it. But, as we know, it did have a problem of another order. Probably the first person to notice it was the English astronomer and mathematician Thomas Digges, in the year 1576.

ANAXIMANDER, ANAXIMENES, ARCHELAUS, XENOFANES, DIOGENES, LEUKIPPUS, DEMOCRITUS AND EPICURUS TEACH THAT THERE IS INFINITE NUMBER OF WORLDS AND THAT THEY ARE SPREAD TO INFINITY IN ALL DIRECTIONS.

Anaximander, A17

The Paradox Discovered

Thomas Digges, a great supporter and fervent propagator of heliocentrism, took Copernicus' planetary system and placed it into the universe of the ancient Epicureans. In so doing he eliminated the sphere of the fixed stars, the last relic of Aristotelianism. The only reason for its existence was common rotation of all fixed stars, and by ending that rotation the sphere became redundant. By scattering the stars into endless space he took the logical step that Copernicus had not taken. Digges therefore assumed an infinite universe with an infinite number of stars. And he soon noticed that something was amiss in that model—it didn't account for the dark sky at night. So Digges discovered an enigma, which we now call the dark night sky paradox or the photometric paradox and sometimes also the Olbers' Paradox.

Why is it dark at night?

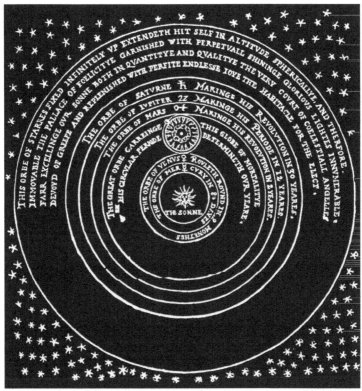

The universe according to Thomas Digges. (This depiction stimulated speculation. Are we to understand it "literally", in the scale? If yes, then the distances of stars are relatively small and our Solar system played an exceptional role in the universe—contrary to the original view of Epicurus. However, to depict the situation in scale is quite impossible. If the Solar system is only few millimetres big, the nearest stars should be several kilometres away.)

THIS ORB OF STARS FIXED INFINITELY UP EXTENDS ITSELF IN ALTITUDE SPHERICALLY, AND THEREFORE IMMOVABLE THE PALACE OF FELICITY GARNISHED WITH PERPETUAL SHINING GLORIOUS LIGHTS INNUMERABLE, FAR EXCELLING OVER (THE) SUN BOTH IN QUANTITY AND QUALITY THE VERY COURT OF CELESTIAL ANGELS, DEVOID OF GRIEF AND REPLENISHED WITH PERFECT ENDLESS JOY, THE HABITACLE FOR THE ELECT.

His claim to have discovered it was somewhat undermined by the fact that he quickly came up with a solution that later turned out to be erroneous. He said the eye was incapable of receiving rays from the most distant of the stars, and so most of the points of light remain invisible to us because they are just too far away.

> . . . EVEN TYLL OUR SIGHTE BEINGE NOT ABLE FARDER TO REACHE OR CONCEYVE, THE GREATEST PART REST BY REASON OF THEIR WOUNDERFULL DISTANCE INVISIBLE UNTO US.

> *In a further text Digges seemed to take fright at what he had done, realising that by abolishing the last sphere he was actually abolishing heaven! He therefore hastily reassured his readers that he had not forgotten about the seat of the Almighty:*
>
> AT THIS (WONDERFUL DISTANCE BEYOND SIGHT) MAY WELL BE THOUGHT OF US TO BE THE GLORIOUS COURT OF THE GREAT GOD . . .
>
> *Digges thus brings to mind the Pythagorean Philolaus, who similarly could not neglect the seat of God (Zeus at that time) and placed it close to the "hearth of the universe", which was also unseeable, beneath the other side of the Earth.*

Digges "solution" to the paradox seemed obvious and natural, except that it contradicted the law of optics that stated that light intensity varies inversely as the square of the distance from the source. In defence of Digges, it must be noted that this particular law of optics was not described by Johannes Kepler until 28 years later. By then Digges had been in his grave nine years.

The Stoic and Aristotelian Universes Reincarnated

Johannes Kepler also realised that the darkness of the night sky begged a profound cosmological question. He touched on the dark night sky paradox in his work *Conversation with Galileo's Sidereal Messenger* (*Dissertatio cum nuncio siderio nuper ad mortales misso a Galilaeo, Matematico Patavino*, or *Conversation with the Sidereal Messenger Dispatched among Mortals by Galileo, Mathematician of Padua*, Prague, 1610). This was his reaction to the recent *Sidereal Messenger* (*Sidereus Nuncius*, Venice 1610), in which Galileo had summarised the results of his first observations using a telescope. In his *Conversation* Kepler writes:

> . . . IF THEY ARE SUNS HAVING THE SAME NATURE AS OUR SUN, WHY DO NOT THESE SUNS COLLECTIVELY OUTDISTANCE OUR SUN IN BRILLIANCE? WHY DO THEY ALL TOGETHER TRANSMIT SO DIM A LIGHT TO THE MOST ACCESSIBLE PLACES?

This is perhaps the main reason why Kepler did not accept the idea of stars being "foreign suns similar to our Sun".

Kepler advanced a further argument:

> SUPPOSE THAT WE TOOK ONLY ONE THOUSAND FIXED STARS, NONE OF THEM LARGER THAN ONE ARC MINUTE (YET THE MAJORITY IN THE CATALOGUE ARE LARGER).[23] IF THESE WERE ALL MERGED IN A SINGLE ROUND SURFACE,

---

[23] One arc minute is roughly the resolution limit of our eyes. (Tycho Brahe was considering about two arc minutes.)

Peter Zamarovský

THEY WOULD EQUAL (AND EVEN SURPASS) THE DIAMETER OF THE SUN...

We can see here, however, that Kepler's argumentation supposed angular diameters of stars, which was greatly overestimated, of course. Later, in his last major work *Epitome of Copernican Astronomy* (*Epitome astronomiae Copernicanae*, 1618) Kepler posits an explanation which is based on the assumption that we do not see the entire universe, only part of it. That part is enclosed in a sort of opaque vault. Kepler thus returned in a sense to the old Aristotelian-Ptolemaic model, i.e. a universe that had some sort of boundary. (On the other hand, Kepler's approach is similar to a later solution that assumed the absorption of light by an interstellar substance.)

The deeply religious Kepler also had purely theological objections to the Epicurean universe. How could God arrange everything for people and how could Man be a divine masterwork if the universe was full of worlds and suns just like our own? And so Kepler, himself a persecuted protestant, opposed the views of the persecuted Giordano Bruno, whose most serious offences included the fact that he considered the universe to be infinite. Kepler described Bruno's opinions as "horrida"—frightful.

IN THE CENTRE (OF THE AREA OF FIXED STARS) THERE IS SURELY A CERTAIN GREAT VOID, AN EMPTY HOLLOW SURROUNDED BY FIXED STARS. IT IS CLOSED AND BOUNDED BY SOMETHING LIKE A WALL OR A DOME. OUR EARTH WITH SUN AND MOVING STARS (PLANETS) IS LOCATED IN THE HEART OF THIS IMMENSE EMPTY CAVITY.

Epinome astronomiae Copernicianae, I/II

> *So Kepler thought that there was an enormous gap between the Sun orbited by its planets and the Milky Way. The Milky Way and the fixed stars enclosed the cosmic space and held the Sun at its centre. And all the stars were roughly the same distance away.*[24] *Neither Copernicus, nor Kepler consistently abandoned an anthropocentric approach.*

So what was the situation of cosmology at the beginning of the seventeenth century? The Aristotelian-Ptolemaic geocentric system was no longer sustainable, while the Epicurean model had run up against the obvious fact of darkness at night. Hence most astronomers returned to the Stoic universe—an island of stars floating in an infinite ocean of empty space. Otto von Guericke (1602-1686) was among those who demonstrated that the Stoics' model could solve the riddle of darkness. In his view, God had created the finite world at the centre of an infinite evacuated space.

But this second incarnation of the Stoics' cosmos was also short-lived. It was Newton himself who buried the concept—although it was a premature burial, as we shall see. According to Newton's law of universal gravitation all bodies attract each other. A finite collection of stars could not remain at rest, but would collapse into their centre.[25] Therefore the universe must be infinite. The planets did not fall into the Sun only because they were revolving. However, our Solar System had an exceptional position. God had imposed no motion on other heavenly bodies.

---

[24] This idea resulted chiefly from the fact that the parallax of the stars had not been measured. See the articles about Aristarchus and Tycho Brahe in the second part of this book.

[25] In his letter to the reverend Richard Bentley (1662-1742)

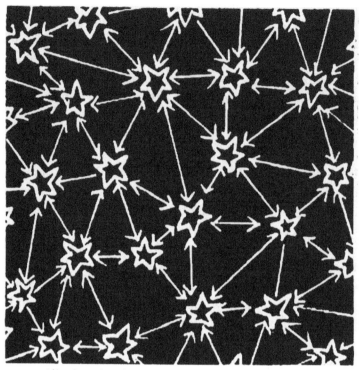

All celestial and terrestrial bodies attract each other.

A portent of the law of universal gravitation can be found in the writings of the Greek philosopher Anaximander of Miletus:

OUR EARTH IS FREELY FLOATING IN SPACE, HELD BY NOTHING, FOR IT IS AT THE SAME DISTANCE FROM EVERYTHING.

A11 from Hippolytus

> *The idea that a finite universe ought to collapse under its own gravity (weight) was already expounded by the late Epicureans. Lucretius Carus wrote:*
>
> IF ALL THE SPACE IN THE UNIVERSE WERE SHUT IN AND CONFINED ON EVERY SIDE BY DEFINITE BOUNDARIES, THE SUPPLY OF MATTER WOULD ALREADY HAVE ACCUMULATED BY ITS OWN WEIGHT AT THE BOTTOM, AND NOTHING COULD HAPPEN UNDER THE DOME OF THE SKY—INDEED, THERE WOULD BE NO SKY AND NO SUNLIGHT, SINCE ALL THE AVAILABLE MATTER WOULD HAVE SETTLED DOWN AND WOULD BE LYING IN A HEAP THROUGH ETERNITY.
>
> Lucretius, *On the nature of things*

After Newton's discoveries everything that had previously had value was devalued: not only the Aristotelian-Ptolemaic cosmos and the Epicurean universe but also the last "reasonable" concept—the Stoic universe. The question of the structure of the universe was back at the very beginning.

But where to start? After all, it was not likely that any entirely original yet still rational approach existed. And so the astronomers decided to recycle the old Epicurean concept of a homogeneous universe. It still seemed the most acceptable model—it did not require any inexplicable or even supernatural acts: no creation and no end of the universe and it did not attribute to us any unwarranted sovereignty. The only problem was the darkness of night sky. And so a search began for ways to supplement and improve on the Epicurean model to avoid the paradox.

In 1720 Edmond Halley (1656-1742) came up with the idea that the paradox would be solved if the intensity of light decreased more rapidly than in proportion to the square of its distance. He presented his solution at a lecture that Newton himself was chairing. Astonishingly enough Newton voiced no objection. Was he asleep? Maybe—he was nearly eighty. But it is also possible that he was reluctant to challenge Halley in public. They were friends and he was indebted to Halley for a great deal. Above all for having funded the first edition of his celebrated *Principia*. According to Halley's calculations light should decrease in proportion to the fourth power of distance from its source. But he had miscalculated by confusing the apparent magnitude and the absolute magnitude of the stars. His mistake was subsequently pointed out by the astronomer Wilhelm Olbers.[26]

In 1744, the Swiss mathematician Jean-Philippe de Chéseaux (1718-1751) undertook a more detailed analysis of the Epicurean model. He calculated that we should be able to see up to a distance of three thousand trillion (i.e. $3 \cdot 10^{15}$) light years! At that moment 10^{46} stars would be shining over our heads, filling the entire sky. The Earth would confront a glow 91,850 times greater than on a sunny day,[27] because the sky occupies a solid angle 91,850 times greater than the solar disc. (Chéseaux assumed that the stars shone with the same intensity as our Sun. This assumption turned out later to be correct.)

> *Chéseaux assumed that the average density of stars was as in our surrounding in the Galaxy. We now know that stars are grouped into galaxies and galaxies are grouped into groups and clusters. Their average concentration is therefore lower and the visibility should even be greater—an estimated 10^{23} light years.*

De Chéseaux came up with his own solution to the darkness at night—we are protected from a fiery hell because the light is swallowed up by interstellar matter.[28] For this to happen, the

---

[26] Halley, Tipler (1988a)
[27] The glow would also be throughout the whole universe, as the Epicurean model assumes homogeneity.
[28] Astronomers talk about interstellar extinction.

interstellar space could have to be perfectly transparent, from the routine standpoint, more transparent than crystal glass or the purest mountain air. For adequate absorption it would need to be $3 \cdot 10^{17}$ times more transparent than pure water!

> *Because of the dimensions of the universe, what seems perfectly transparent on an earthly scale can be quite opaque on a cosmic scale. If the universe were filled with air—as was tacitly assumed in the past—we would not even see our nearest planetary neighbour, the Moon.*

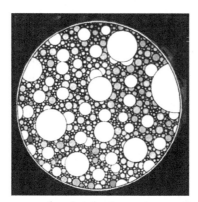

Remote stars appear less bright due to interstellar absorption

Let us go back to our analogy with the forest. We are standing in the middle of a forest and can see trees in all directions. The average visibility is 50 metres, for instance, and we can see roughly 2000 trees. But then a mist—i.e. an absorbent medium—comes down and we are only able to see the hundred nearest trees up to ten metres away, and between them fog. And that fog represents cosmic darkness. After de Chéseaux's calculations, scientists were placated—the problem of the dark night sky had finally been solved: the universe was infinite, eternal and homogeneous, but between the stars there lay absorbent medium, either gas or some dust or other. It veiled the very distant stars and shielded us from their light and heat, in the same way that sea water prevents the Sun's rays reaching the bottom of the ocean.

The solution to darkness at night based on the idea of interstellar light absorption soon won favour. It was also supported by the distinguished astronomer Wilhelm Olbers.

> *Nowadays the dark night sky paradox is often called the Olber's paradox, a name given to it by the cosmologist Hermann Bondi, who ascribed the paradox to the astronomer in his monograph from 1952. But Bondi was a bit hasty. Olbers didn't discover the paradox nor did he find a correct solution. His contribution consisted solely in popularising it in a series of articles between 1823 and 1826.*

An interesting variation on the absorption solution was proposed by the Irish astronomer Edward Fournier d'Albe (1868-1933). In his view the darkness between the visible stars was created by invisible dark stars which were in the majority. A bright star glimmers through them here and there. If that were truly the case, the stars would have to keep going on and off due to their changing alignment.[29] Fournier's universe with permanently appearing and disappearing stars would look quite different from the one we inhabit. However, it would seem that Fournier himself did not take this solution too seriously.

The fatal blow to the absorption theory was dealt by John Herschel (1792-1871) in 1848. He argued that in an eternal universe the temperature would have to equalise. Incandescent stars and cold absorbent matter could not coexist for ever. Matter itself would have to incandesce and glow or the stars would have to cool down and stop shining.[30] If the universe was eternal then the light of the stars could not be engulfed to any great degree by interstellar matter.

---

[29] The stars' own (proper) motion had already been recorded by Edmond Halley in 1718. (See later.)

[30] It was not until the 1850s that the notion of light as a carrier of energy gained general acceptance. Not even Herschel was aware of it at the outset. (Tipler 1988a)

Although light absorption does not solve the enigma of darkness, it does play an important role locally in the universe. In many parts of the sky opaque interstellar clouds can be observed— dark dust nebulae.[31] They are also spread in the Sagittarius constellation and prevent us from seeing the nucleus of our own galaxy with its enormous black hole and many densely clustered stars of great interest. Were it not for these dark nebulae the core of the galaxy would shine more brightly than the Moon when it is full.[32] This begs the question: "How is it possible that light absorption works locally but not globally?"

The reason is that absorption is able to protect us from heat of the stars only if the absorbent matter does not itself incandesce. If there were lots of absorbent matter, as there would have to be in order to explain the darkness of night sky (since the sky is dark for the most part) the temperature in the universe would equalise in time. But if there is not much absorbent matter (the local case), the temperature does not equalise because the absorbed heat is radiated to the empty dark space (mostly in the form of invisible infra-red radiation). Put simply, for absorption to work there must not be too many stars or absorbers, and on the contrary a lot a dark empty space has to remain.

---

[31] It would be more apt to refer to them as smoke, because dust results from the pulverisation of some solid matter (rock), whereas smoke comes from condensation of a gaseous phase.

[32] An odd source of radio waves in the Sagittarius constellation was recorded as early as the 1930s in the radio map of Karl Jansky and Grote Reber. It lies in the direction where the core of our Galaxy ought to be located 25,000 light years away. (The dust clouds seem transparent for infra-red radiation and radio waves.)

> *Based on the assumption that the temperature of various parts of the universe was equalising and mechanical energy was gradually being changed into heat,[33] the German physicist Hermann von Helmholtz (1821-1894) and the British scientist Lord Kelvin (1824-1907) deduced a further remarkable paradox: the universe ought to move not only towards a state of thermal equilibrium but also towards thermodynamic equilibrium (a state of maximum entropy). But this would imply that cosmic evolution as a whole must come to a halt.[34] All energy is totally transformed into heat, i.e. microscopic movement of atoms and molecules. The universe will have the same temperature throughout[35] and all macroscopic movement will stop. The universe will undergo thermal death—life will be out of the question. So both paradoxes—the dark night sky and the absence of thermal death of the universe—derive from the same question: How is it possible that the universe is in a state of (thermodynamic) disequilibrium?*

So we are back with the problem of dark sky. And apart from the "rational" explanations there are some that a more of a joke. They include Fournier's later solution proposed in 1907, namely, that most of the stars are invisible because they are ranged one after another like soldiers on parade. The absurdity of this solution is immediately obvious: not even the omnipotent God could have choreographed their alignment, in view of the stars own motion.

---

[33] A process known as energy dissipation.
[34] The conclusion follows from the second law of thermodynamics (the law of increasing entropy) and the assumption that the universe can be conceived as a closed system.
[35] This temperature would have to be few degrees above absolute zero, so it would essentially be total freezing. In this situation total darkness would reign in the universe.

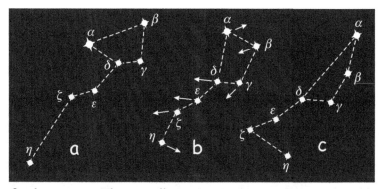

Even fixed stars move. The constellation Big Dipper 100,000 years ago (a), now (b), and 100,000 years from now (c).

In terms of the human life span, the movements of the stars are negligible. The sky appears unchanged and astronomers are not obliged to update their maps all the time. So how can one measure such tiny movements at all? Halley did it by extending the time base. He compared the present locations of stars with their locations in Ptolemy's catalogue from the second century CE, which also contained earlier data from Hipparchus (c. 190-125 BCE). He also used data from Tycho Brahe (1546-1601) and John Flamsteed (1646-1719). He noted a change of position in the case of the stars Arcturus, Sirius, Procyon and Palilicius (the former name of the star Aldebaran). What had Halley discovered is the so-called proper motion of stars. This is defined as the angular change in their apparent position (relative to the "fixed stars") over time as seen from the Earth.[36] And so thanks to Halley, the fixed stars were no longer fixed. It took a long time for astronomers to appreciate the implications of that crucial discovery. It is evident, therefore, that the stars are not hiding behind each other—or only exceptionally, in statistically insignificant numbers (mostly in the case of double stars in close proximity). However, another geometrical solution might seem more reasonable. The stars might group together in

---

[36] Proper motion has to be distinguished from the daily motion caused by rotation of the Earth. Now we define the proper motion more precisely as the motion seen from the centre of mass of the Solar System.

a hierarchical fashion into larger and larger clusters. They could go on endlessly creating galaxies, galaxies of galaxies, etc. As space grew in scale the average concentration of stars would diminish and this might solve also the riddle of the dark night sky.

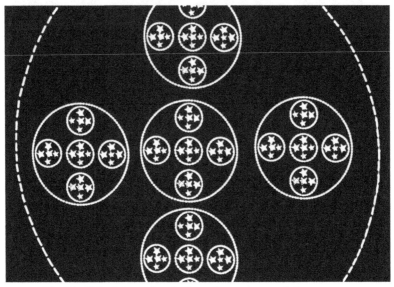

Model of the hierarchical (fractal) structure of the universe

In this case the number of stars in the individual layers of our "onion model" would grow more slowly than the volume of each layer and the average density of the universe would fall to zero as it grew in scale. The remoter layers would shine less and less and the overall sum of light from all the layers could have a finite value. If, for example, each further layer shone with half the brightness of the previous one, the total radiance of the sky would be twice the radiance of the first layer, because as we know from mathematics, the infinite series:

$$1 + 1/2 + 1/4 + 1/8 + 1/16 + \cdots = 2$$

Hierarchical cosmology was proposed in 1848 by John Herschel and it soon achieved great popularity. It was elaborated in 1908 by Carl Charlier, who drew inspiration from Fournier's study of 1907.

Benoît Mandelbrot (1924-2010) returned to the idea in the 1970s and explained the darkness of the night sky by suggesting that the stars were fractally distributed. This was a hierarchical arrangement as described by Charlier: the structure of the star clusters would be repeated on a greater and greater scale and as they grew in scale the average concentration of stars would fall to zero.

> *A fractal is the name given to a geometrical object that is scale-invariant, in other words, irrespective of the distance from which it is observed (or on what scale) a fractal always appears the same (it is self-identical) or at least similar (self-similar). The shape may appear extremely complex at first sight, but from a mathematical point of view it is generated by repeated use of a simple formula. Fractals are to be found among many natural features, such as lunar craters, mountains, clouds, shores, rivers, the vascular system, etc. They are also common in the plant kingdom: one attractive and tasty example is Romanesco broccoli or Roman cauliflower. But we are straying away from astronomy into gastronomy.*

There are many self-similar fractal structures to be found in the universe. Stars tend to combine into groups of two or more. Our own Sun is also part of a free grouping of stars which travel together through space at about the same velocity and in the same direction. Stars also combine into gigantic galaxies. And those galaxies in turn group themselves into clusters of galaxies, and then into superclusters. Does this hierarchical structure continue truly without end? As late as the 1970s the indications were that it did. But astronomical equipment went on developing and new observations suggested ever more strongly that higher structures lacked a hierarchical character. Instead they were more like some kind of cell and on a scale of millions of megaparsecs[37] the universe is homogeneous.

---

[37] A parsec is a unit of distance equivalent to about three light years. It is the distance from which we would be able to see the radius (more exactly the large half-axis) of the Earth's orbit around the Sun at an angle of one arcsecond.

> *When we talk about the distribution of matter in the universe the tacit assumption is usually that it consists of shining stars. But recently astronomers discovered that only about one per cent of the matter emits light. A large part of it is dust, gas and non-luminous bodies. In addition there is a considerable amount of so-called dark matter in the universe, which is in fact not dark but perfectly translucent and therefore invisible. Its physical nature is still totally unknown.*

Where to from here?

We have now looked at many attempts to solve the mystery of the dark night sky and have seen how each one in turn failed. What might have succeeded in terms of geometry failed from the point of view of physics, and what might have satisfied the demands of geometry and physics was inconsistent with observation. The story of the paradox therefore continues.

Paradoxes and contradictions are not located in nature, however, but in our approaches to nature—in our heads. We simply fail to understand the universe and our lack of understanding leads to controversy. But where do we go wrong? Astronomers keep going back to old outdated models in search of what might have been overlooked. And thus at the turn of the twentieth century the Stoic notion of an infinite universe, in which matter (i.e. the luminous stars) is located only in one limited part of space, underwent its third reincarnation. This time the island of stars was identified with our own Galaxy, the Milky Way. Astronomers realised that they had rejected the Stoics' model of the universe prematurely. Newton's law of universal gravitation does not actually rule out such a universe. But if the Stoics' model is to be a representation of the actual universe, i.e. one in which the stars attract each other, then the universe cannot be static. The stars have to move. Previously that was tacitly ruled out. Even though Halley had already noted the movement of several stars at time of Newton his discovery was not fully appreciated and was not receive general acceptance. At the beginning of the 20th century, however, the situation had changed and no one doubted any longer that the stars themselves move. Their motion was due to inertia and inertia prevented motion—which might well be described as free fall—stopping at any point. The stellar system goes on collapsing but doesn't collapse altogether because the stars do not collide in their flight. Because of the enormous distances between them collisions are extremely rare.[38]

> *Our terrestrial experience tends to be misleading when viewed in cosmic terms. This also applies to the notion of "collapse". There is an enormous difference between collapse on a terrestrial scale and collapse on a cosmological scale. Not only Lucretius failed to realise that "cosmic collapse" was something different, but so too did the astronomers of the Renaissance. These fundamental differences are best described in the language of thermodynamics. When a house collapses, for instance, the potential (stored) energy of its bricks escapes in the form of released heat. There is no going back; the fall is irreversible: you can't rebuild the house by heating up the bricks. But in the stellar system the events of the fall are reversible except in the very rare case of an actual hard collision. The potential energy of a falling star is transformed into its kinetic energy, then back into potential energy, and the process is repeated over and over again. No potential energy is transformed into heat. And so it can all be repeated for ever (or for a very long time, at least). The difference is due precisely to the extreme dilution of cosmic matter and hence the absence (or extremely low intensity) of so-called dissipative processes, i.e. those which transform mechanical (i.e. potential and kinetic) energy into heat.*

The conditions in the universe can be compared to putting five table tennis balls inside a hollow cavity the size of the Earth, in which the balls move at a speed of several millimetres per hour. It is very unlikely

---

[38] Or more precisely, because of the high ratio between the average distance between stars and the average size of the stars.

we would live to see any of them collide. That is why stellar systems survive in a state of dynamic equilibrium—a process of "collapse"—for millions and billions of years. There are also many systems that rotate as an entity, such as spiral galaxies, including our own galaxy with a capital G. On a smaller scale a similar situation also applies to our Solar System, in which centrifugal forces prevent the planets (and some comets and asteroids) from falling into the Sun.

> *There are hundreds of billions of stars in our Galaxy but direct collisions between them are extremely rare. Were stars to actually collide, however, it would result in a magnificent firework display—to be viewed at a safe distance, of course (otherwise we'd risk the tragic fate of Pliny the Elder, who set out to investigate the eruption of Vesuvius in 79 CE). In the space of several minutes there would be a sudden increase of temperature and pressure in the stars. The thermonuclear reaction would spread throughout the whole volume of the star, resulting in a mighty explosion. No astronomer has yet been lucky enough to see this rare phenomenon, so stellar collisions are only studied in theory.*
>
> *This is not the case of galactic collisions, however. Unlike the stars, the distances between galaxies are comparable to their size and so collisions between galaxies are frequent, with one galaxy penetrating another for millions of years. (It sometimes ends with one galaxy swallowing another, a process described as "galactic cannibalism".) The stars themselves almost never collide during this process, instead they only exert gravitational influence on each other. When galaxies penetrate each other, however, enormous clouds of interstellar gas condense and this process accelerates the creation of new stars. If you hang around for three billion years you'll see our Galaxy collide with galaxy M31 in the Andromeda constellation. That collision will lead to the creation not only of many stars but also of many planets like ours, planets inhabited by life, perhaps also by creatures similar to us. Life in our vicinity will then be a lot more varied and more plentiful.*

The reborn Stoic model, namely the idea that all stars are part of our Galaxy, which slowly rotates in empty space, was still advocated in 1920 by the American astronomer Harlow Shapley (1885-1972). The Stoic model's third incarnation was even shorter than the previous ones, however. Its fate was sealed by rapid developments in astronomical instruments. Ever larger and more powerful telescopes were being built and astronomers were obtaining photographs of the remote reaches of the universe of ever greater quality. Spectroscopy went on to differentiate the radiance of stellar systems (galaxies) from the radiance of gas nebulae.[39] By 1923 Edwin Hubble had discovered that much of what were considered nebulae did not consist simply of trifling clouds adorning the fringes of our native Galaxy. Nor were they relatively small star clusters. On the contrary, they were enormous galaxies similar to our own. And as telescopes became ever more powerful more and more foreign galaxies were detected. Billions of galaxies. And our native Galaxy is no more than an insignificant little speck of dust in an enormous universe.

The universe according to Shapley consists only of our Galaxy with peripheral star clusters and nebulas.

---

[39] Stars have essentially a continuous spectrum, while the light emitted by dilute gas has a line spectrum.

> *Shapley's mistake was due to a serious error in calculating the distance of the M31 galaxy. Because the estimate was too low, it was thought to be an object on the fringes of our Milky Way. The outdated view that all the "clouds" that we see in the sky are nebulae still survives in astronomical nomenclature. The M31 galaxy is still sometimes referred to as the "Andromeda Nebula". Determining the distance of cosmic objects is fundamental for understanding the structure of the universe. It is a problem that has dogged the steps of astronomy since it split off from astrology. (The distance of cosmic bodies is not of importance in astrology, only their apparent position in the sky.) And in spite of the enormous progress made in all fields of astronomy, the determination of cosmic distances is still a crucial task. Out of the various methods used for measuring cosmic distances, the parallax method has played a key role, but it can only be used for the nearest stars. For more distant objects the "standard candle" method is used. This involves looking for types of stars whose absolute luminosity is known ("standard candles") and then determining their distance according to their apparent luminosity.[40] Distances are also determined indirectly from the so-called redshift. This method is only applicable to truly "cosmological" distances, i.e. remote galaxies. (It cannot be used for the M31 galaxy, for instance. Two million light years is not far enough away in "cosmological" terms.)*

There are estimated to be up to one trillion galaxies within a distance of 10 billion light years. But will the "forest of galaxies" really stretch to infinity? And if it does, how come it is dark at night?

---

[40] For that purpose variable stars (Cepheid Variables) are used, and for greater distances type Ia supernovae.

Ether voids

A curious explanation of the darkness at night was formulated by Simon Newcomb and John Gore.[41] It was based on the theory widely spread in 19th century that light needs some medium for its propagation. The solution assumed that the medium—ether—was not omnipresent but formed some kind of mighty drops, ether voids. The light that reaches the walls of the drop is reflected back and the space in-between, where there is no ether, is opaque. So the visible universe is merely the interior of our ether area, the drop in which our Galaxy is located.

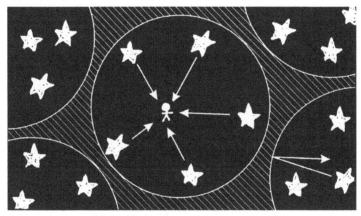

Great "drops" of ether in opaque cosmic space

However, closer analysis showed that if the walls of the ether drops were reflective the stars would be many times more visible. The paradox with dark night therefore remained unsolved. (If the light was not reflected the "drop solution" would resemble the absorption solution or the ancient concept of the last sphere. But it raised the issue of where the energy of the absorbed light was lost.) Lord Kelvin exposed the implausibility of the ether voids theory in 1902.

---

[41] Simon Newcomb (1835-1909) was a Canadian astronomer, John Gore (1845-1910) was an Irish amateur astronomer. They were inspired by the ideas of the Scots engineer and physicist William Rankin (1820-1872).

Moreover, it was obvious from the 1920s that other galaxies were visible.

> *The term "ether" (aether) dates back to ancient times. In Antiquity it was the name given to the hypothetical building material of the celestial bodies.[42] However, the decline of Aristotelianism also meant the decline of ether. Nevertheless in the seventeenth century ether made a comeback. This latter-day ether was no longer the material of the celestial bodies but the hypothetical liquid substance that filled space itself, a medium that enabled the propagation of light. Reborn ether did not have a very long lifespan, however. It was already called into doubt in 1801 by the experiments of Thomas Young. Studies of the polarisation of light showed that light waves were transverse, unlike sound waves, which are longitudinal.[43] And as physics tells us, transverse waves cannot pass through liquids. Therefore ether could not be a liquid, but would have to have the structure of an elastic solid. Modelling such a structure involved many difficulties. Experiments performed by Albert Michelson and Edward Morley in 1881 and 1887 finally put paid to the ether theory. These experiments demonstrated that light moves in all directions at the same speed regardless of the movement of the hypothetical ether.*

---

[42] "Ether" was translated into Latin as the "quinta essentia", or the quintessence, i.e. the "fifth essence". (However, this name was not used in modern times to describe light-carrying substance.) Recently the term "quintessence" emerged afresh in cosmology, but denoted something else.

[43] Polarisation (linear polarisation) of light refers to the phenomenon when only waves that oscillate in a single plane are selected out of waves of differing planes of oscillation. Therefore only transverse waves can be polarised.

The End of the Eternal Universe

After these failures astronomers realised that more fundamental steps had to be taken to explain the riddle of darkness at night. It was necessary to jettison some unconscious assumptions and age-old preconceptions. Above all it meant abandoning the static viewpoint that had been a typical feature of all the old models. Since time immemorial people had been enthralled by the stars because of their immutable nature, which was so in contrast with the ephemeral world of us mortal beings. The fixed stars, as the name implies, were static and, moreover, they hung in the celestial sphere, which turned uniformly with them. Later the sphere stopped and the Earth itself started to rotate. But this only served to confirm still more the static nature of the realm of stars. And when astronomers got rid of the last sphere and the stars were scattered freely throughout space, the stars still remained essentially fixed. Halley's discovery of the proper motion of stars concerned only five of them, and moreover this motion was very slow. Astronomers ignored it. It was even as if the stars did not have their own evolution, their own "life" or "internal movement". Not even the development of the whole universe attracted speculation. The static view was firmly anchored not only in people's consciousness but above all in their unconscious. It constituted a collective prejudice that blocked any forward progress.

UNLESS YOU EXPECT THE UNEXPECTED YOU WILL NEVER FIND IT.

Heraclitus, B18

The alternative view, namely the idea that stars evolve, that they come into existence and expire, was regarded as peripheral speculation in early modern times. At the same time the view that the universe as a whole once came into existence (was created), that it evolves and at some point in time will cease to exist, tended to be regarded as just another outdated religious notion. Nevertheless the model of an evolving universe has a venerable tradition as a part of philosophy and also as a part of what might be termed science. Two thousand years ago, the ancient Greek Atomists were already teaching about the evolution of individual "worlds"—"cosmoses". And the Stoics (like the Hindus in the East) were speculating about how the universe emerged out of fire and after a certain period of time it would end in fire (*ekpyrosis*).

The preconception of an eternal static universe was not overcome until the mid 19th century. And when it was overcome its connection with the dark night paradox immediately became apparent—thanks to an astronomer and a poet.

Edgar Allan Poe and the darkness at night

It is not often that a poet helps solve a scientific problem. So when someone asserted recently that Edgar Allan Poe had helped explain the dark night sky paradox it attracted attention. It was the American cosmologist Edward Harrison who gave him credit for this. Was he right?[44]

So let us take a look at *Eureka—A Prose Poem*, in which Poe declares that he intends to speak "of the material and spiritual universe". Concerning the riddle of the dark night he says:

WERE THE SUCCESSION OF STARS ENDLESS, THEN THE BACKGROUND OF THE SKY WOULD PRESENT US A UNIFORM LUMINOSITY, LIKE THAT DISPLAYED BY THE GALAXY—SINCE THERE COULD BE ABSOLUTELY NO POINT, IN ALL THAT BACKGROUND, AT WHICH WOULD NOT EXIST A STAR. THE ONLY MODE, THEREFORE, IN

---

[44] Those who think this include the American physicist Frank Tipler (1988a, b)

> WHICH, UNDER SUCH A STATE OF AFFAIRS, WE COULD COMPREHEND THE VOIDS WHICH OUR TELESCOPES FIND IN INNUMERABLE DIRECTIONS, WOULD BE BY SUPPOSING THE DISTANCE OF THE INVISIBLE BACKGROUND SO IMMENSE THAT NO RAY FROM IT HAS YET BEEN ABLE TO REACH US AT ALL.

Thus Poe assumed that the universe came into existence (was created) at some moment in the past. In the span of time since the beginning of the universe, a ray of light could only travel a finite distance.[45] So everything what we see in the universe is the past. But light has not had time to reach us from the remotest corners of the universe, so we can only see the nearest stars, and there are not enough of them to illuminate the entire sky. The furthest distance we can see is equal to the product of the age of the universe **t** and the speed of light **c**. Therefore the only stars that are visible are within a sphere with the radius **c × t**.

> *For instance, we observe the Moon as it was a second ago and we see the Sun with a delay of eight minutes. We see Sirius as it was when we were eight years younger and star cluster Pleiades as it was at the time when Elizabeth I was Queen of England and Tycho Brahe and Johannes Kepler were wandering about Prague Castle with the emperor Rudolf II. Such delays are not important in terms of the history of the universe. The situation starts to be of more interest when we enter the "deep sky". The now familiar galaxy M31—"the large nebula in the Andromeda"—appears to us as it was almost three million years ago, i.e. at the time when the first hominids started to appear on Earth. And this galaxy is one of the nearest—we can glimpse it even with the naked eye. When we look into even more remote corners of the cosmos a truly deep past opens up to us. There we observe cosmic objects that ceased to exist long before our Solar System and our*

---

[45] The speed of light had been measured by the Dane Olaf Rømer in 1676. (See Part II)

> *Galaxy were born. In this way the most spectacular historical film is playing before the gaze of astronomers—the history of the entire cosmos.*

Awareness of the time dimension of the visual image breached the traditional perception of the universe. It also rehabilitated the idea that the universe had a beginning. Although the notion that the universe came into existence (was created) formed a part of the oldest myths, scientific circles in the nineteenth century regarded it as outdated religious prejudice.

> *The creation of the universe begs the question: what was it created out of. Within the various mythological and religious systems there are essentially two explanations of the origin of the universe. The first assumes the emergence of the cosmos (i.e. an ordered world) from primordial chaos. This was also espoused by many philosophers, including Empedocles, Anaxagoras and Plato. Akin to this is the emergence of the cosmos from a primordial fire, a notion held by the Stoics. The second explanation assumes the creation of the world out of nothing. That is how the Biblical message is traditionally interpreted. However, the Book of Genesis does not specify from what Earth and Heaven[46] were created.*

---

[46] Old Hebrew had no term for cosmos.

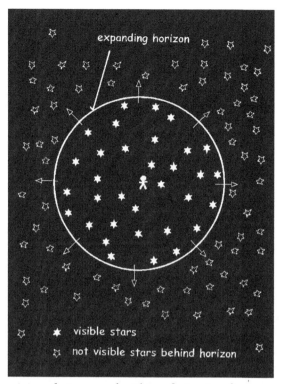

Fig. 25 In a universe that emerged within a finite time frame we can see only the nearest stars. These are enclosed within a sphere (horizon) whose radius is expressed by the formula $r = c \times t$, where c is the speed of light and t the age of the universe.

Although Poe's explanation turned out to be incomplete his idea of a "dynamic approach" played a fundamental role not only in solving the paradox but also in opening an entirely new perspective. Cosmology—the study of the structure of the universe—finally linked up with cosmogony, the study of the origin and evolution of the universe.

Soon arguments against Poe's precedence began to appear. In particular the assertion that at the time *Eureka* was written, the "dynamic" solution was already known about.[47] It was said to have

---

[47] Tipler (1988a, b)

been already proposed by the German astronomer Johann Mädler in his popular books on astronomy. In 1858 Mädler really wrote:

THE WORLD IS CREATED AND HENCE IT IS NOT ETERNAL . . . IN THE FINITE AMOUNT OF TIME IT COULD TRAVEL BEFORE IT REACHES OUR EYE, A LIGHT BEAM COULD PASS THROUGH ONLY A FINITE SPACE NO MATTER HOW GREAT THE SPEED OF LIGHT. IF WE KNEW THE MOMENT OF CREATION, WE WOULD BE ABLE TO CALCULATE ITS BOUNDARY.

Mädler 1858

Mädler's solution to the dark night paradox based on the finite age of the universe was thought highly of.[48] So had Poe borrowed his solution of the paradox from Mädler? In *Eureka* Poe indeed mentions Mädler, but not in connection with the solution to the darkness at night—and it is not cited exactly; after all, it was "only a poem". But the most compelling counter-argument is that in Mädler's *Popular Astronomy* of 1841, which Poe could conceivably have read, an entirely different solution to the paradox was advanced: darkness was the result of the absorption of light by an interstellar medium. That solution was advocated even in the book's fourth edition which came out in 1852, four years after Poe wrote *Eureka*. Poe's solution was published by Mädler in 1858 and was included in the entirely reworked fifth edition of his book, dated 1869—twenty years after Poe's death.

Another objection to ascribing the solution to Poe is the non-standard way that he presented it. *Eureka* is certainly no

---

[48] Even the philosopher Friedrich Engels praised it in his *Dialectics of Nature*. He did not even mind the notion that the universe had a beginning, i.e. it was created, which the Marxists would subsequently inveigh against.

scientific treatise. (It is more reminiscent of the didactic epics of Antiquity, in particular the epic of Empedocles in which he too links acute physical insights with dark metaphysics.) Poe was fully aware of being an amateur in matters of science and he issued the caveat: *". . . it is as a poem only that I wish this work to be judged after I am dead."* The non-scientific nature of the text is also emphasised by its parodical references: Euclid is renamed Tuclid and Aristotle is referred to as "the Turkish philosopher called Aries and surnamed Tottle", a deep insult to any true-blue Greek. Nevertheless, the element of parody in the text is not prominent enough, in my view, for it to have been the reason why Poe wrote it. The author himself states that his motive was *"the Beauty that abounds in its Truth; constituting it true"*. Beauty and truth, no less. In spite of the confusing elements that Poe could permit himself as a poet, I believe that he was entirely serious about the ideas he advanced. The fact that the poem was dedicated to the natural scientist Alexander von Humboldt gives further credence to my view.

> *Cosmologists call the furthest distance from which light (or some other signal) reaches us "the horizon". Hence the horizon marks the boundary of observable space—the sphere of the visible. Present-day cosmologists have introduced further types of horizon and refer to the boundary of observable space as the particle horizon. Its radius r increases at the speed of light, because the t parameter changes—the universe is aging.*

The "forest model" that we based ourselves on when describing the paradox therefore has a crucial weakness. It assumes that we see all the trees—stars—as they are at the moment of observation. This assumption is all right for terrestrial forest as light overcomes terrestrial distances almost immediately. But when there are great cosmic distances it literally dawdles, so the situation concerning the visibility of the celestial forest is quite different. But let us return to Poe's and Mädler's notion. As the universe ages the sphere of the visible expands, so the number of stars within that sphere should also increase. More and more light from stars should appear in the sky and the entire universe should gradually light up and be filled with photons arriving from ever more remote places. The temperature should also rise and in time a situation would arise in which the heat would kill life on Earth before burning and evaporating the soil of our native planet. As we shall shortly see, present-day cosmology does not accept this scenario of cosmic evolution. Equally untenable is the assumption that the universe originated precisely as it is now. However, the perception that the darkness of the night sky is bound up with the history of the universe was an important step in the right direction.

Astrophysics Arrives on the Scene

There is no point in studying the universe without addressing the question of its material basis, in other words, what material it is made of and how that material behaves. That is why modern-day astronomy cannot do without astrophysics, nor astrophysics without physics.

For a long time the material aspect of the universe was a mystery. The prevailing perception of the universe in Antiquity and in the Middle Ages was of a superlunary region made of ether, i.e. a material totally different from the material of the sublunary—terrestrial—region. The preconception of a fundamental difference between the superlunary and sublunary worlds prevented consideration of the stars in terms of physics, or more precisely, using the concepts of sublunary, i.e. terrestrial, physics. But even in Antiquity alternative opinions sometimes emerged. Anaxagoras of Clasomenae, for instance, declared that the Moon was made of the same material as the Earth and that the Sun was simply an incandescent rock.[49] The Atomists also believed that parts of the universe, the individual "worlds", were of a similar nature to our own world.

However, the first irreversible (and one might say scientific) step across the gulf between the Universe and the Earth was Newton's discovery of general gravitation, when he realised that the force that drew the famous apple to the ground was essentially the same force that held the Moon in its orbit around the Earth and the Earth in its orbit around the Sun. The next epoch-making occurrence was the invention of spectroscopy, which enables the chemical composition of celestial bodies to be studied by analysing their radiation. The findings were surprisingly unspectacular: celestial

---

[49] In that period the fall of a large meteorite (bolide) was observed. It shone so brightly that it was visible in the sky during the day and its remains were found in a stream not far from Athens. It was a large black rock.

bodies are composed of essentially the same elements as Earth. (So old Anaxagoras was basically right.)

The unification of the physical perspective was a slow process however. It was not until the beginning of the twentieth century that Lord Kelvin showed that the law of the conservation of energy need not apply only to terrestrial processes but also to extraterrestrial events. So Mädler was not mistaken: the stars cannot shine for ever. They radiate energy and consume fuel in the process, but they clearly do not replenish their energy and so it must eventually be exhausted. That was already obvious at the beginning of the twentieth century, though no one had any idea what fuel was used by the stars and how it was burnt.

> *It's fairly understandable that the first notions about the source of energy of the stars were based on terrestrial experience. According to these notions the light from the stars was like the light from a torch, a candle or an oil lamp. It was a combustion process, an oxidation in the course of which heat and light are released. However, terrestrial experience tends to be deceptive when applied to the universe. The idea of "chemical combustion" was called into doubt by John Herschel. He concluded that the combustion would not last long. And Hermann von Helmholtz calculated that our Sun would burn up within 3021 years. Chemistry failed, so it was now the turn of physics. A theory was advanced that assumed that energy was released as stars shrank and the potential gravitational energy was transformed into heat. And this theory in turn was scuppered by the calculations of Helmholtz and Kelvin, which proved that this source of energy would also soon be exhausted.[50] Another hypothesis involved so-called accretions, which were massive falls of interstellar matter onto the surface of stars. It was thought that due to the attraction of the star the falling matter is accelerated to a very high velocity and its kinetic energy is transformed into heat. One advocate of this theory was Robert Mayer (1814-1878), physician and physicist, best known for enunciating the law of the conservation of energy. But the accretion*

> *hypothesis also proved to be short-lived. This was because the falling particles would have to have a considerable mass, otherwise the pressure of the star's radiation would blow them away before they fell. Hence gas and dust, which are the main components of the interstellar matter could not contribute to the influx of energy. And then again, there are not enough larger bodies in interstellar space.[51]*

The crucial shift in opinion did not occur until the mid-20th century, when the astrophysicists included nuclear physics in their arsenal. At that time the East-West power struggle posed a mortal threat to humankind. For science, however, it proved to be a blessing, pushing forward the boundaries especially of nuclear physics and technology. And so thanks to the Cold War we have not just nuclear weapons and nuclear power stations, but also a profound knowledge of the physical processes in atomic nuclei. Nuclear physics subsequently had a decisive impact on astrophysics and cosmology. Understanding the microcosmos was a key to understanding the megacosmos—the universe as a whole. No physicist is in any doubt nowadays that what heats the interiors of stars are nuclear reactions in the course of which nuclei of one chemical element are transformed into nuclei of another element. On that basis physicists have elaborated the scenario of the formation, evolution and demise of stars. They continuously compare their theoretical models with observations of stars at various stages of development.

So how are stars born, how do they live and how do they die? Present-day astrophysics has a clear overall idea: Stars are formed by the condensation of gas nebulae, the basic material of which is

---

[50] Nevertheless heating by means of gravitational compression plays an important role in astrophysics. For instance, it heats up protostars (embryonic new stars) and also so-called brown dwarfs. These are small stars with a relatively low temperature that shine chiefly in the infra-red field. (So they are not brown. No heated body can emit a brown glow.)

[51] Accretion does play a considerable role in the formation of stars, however.

hydrogen. The enormous clouds of hydrogen are unstable structures. It simply requires some external trigger (a collision of clouds, a shock wave produced by the merger of galaxies, the effect of a magnetic field) or their own gravitational attraction for dense gas globules of gas—an embryonic star—to be created. More and more material deposits itself on them from the surroundings (accretion). Then gravitational collapse occurs, and a so-called proto-star comes into being. The proto-star is not yet very hot and it glows partly in the infrared spectrum and partly as visible light. It has a brief existence—just hundreds of thousands or millions of years. The energy released as the clouds fall serves to compress and heat up the material of future star. When the temperature in its core exceeds a million degrees, thermonuclear synthesis is ignited. Hydrogen nuclei—protons—start to fuse and form helium nuclei (two protons and two neutrons), releasing a large amount of energy that heats the globules still more. A glowing star is being born.

> *Once more, our terrestrial experience is of no use to us. What applies on a small scale does not apply on a large scale and vice versa. A small cloud of hydrogen, such as when a fairground balloon bursts, is also unstable and it disperses, but it does not give birth to a star. On a massive scale, however, the situation is different and the force of gravity is decisive. As in the case of steam, drops can condense due to the effect of intermolecular forces, stars are formed from sufficiently large hydrogen clouds through their own gravitational attraction. Like drops, stars are also not formed in isolation. They form groups (associations) of sister stars and sometimes more compact formations—star clusters.*

A star's future depends primarily on its mass. If the star is small and light,[52] it can expect a long and tedious life lasting tens, hundreds and possibly thousands of billions of years. When it has burnt all its fuel it will start to cool and become a so-called white dwarf. This continues to cool and turns into a so-called black dwarf. But

---

[52] But with a mass greater than one tenth of the mass of Sun.

something far more dramatic could also occur. This is because a white dwarf is potentially a thermonuclear bomb of such a size that it could easily wipe out the entire Solar System and everything in the vicinity. If matter from a nearby star accretes onto a dwarf so that its mass exceeds a certain limit, the dwarf becomes alarmed and explodes. A large amount of energy is released and shines like a billion Suns. Any life in the vicinity is totally incinerated. Astronomers refer to such explosions as type Ia supernovae.

Like obese human beings, very heavy stars have shorter lifespans. (Examples of such stellar colossi are Betelgeuse in the constellation of Orion or Aldebaran in Taurus. Astrophysicists expect Betelgeuse to explode soon.) They end in a mighty explosion—a sort of self-immolation. What remains of them is "dust"—neutrons, which, through gravity, form small but very dense neutron stars known as pulsars. Astronomers describe such explosions as supernovae—collapsars. Heavier chemical elements are formed there as products of nuclear combustion (fusion). These are the building material of the planets, the Earth and our mortal frames. (By "heavier elements" astronomers mean chemical elements such as carbon, oxygen, nitrogen, silicon, calcium, iron, gold, etc.) We human beings owe the existence of our bodies to a collapsar of that type. It is unlikely that any god would have managed to knock together the Earth, animals, plants or human beings from just hydrogen and helium.[53] And the most gargantuan of the stars bury themselves alive. Gravity overcomes all the forces that hold their material in its volume. The stars collapse into themselves leaving only black holes behind. It was originally thought that black holes ought not to be visible as they are not luminous, but Stephen Hawking came up with a theory that black holes ought to emit radiation. In the case of bigger holes this radiation would be very weak. The smaller black holes could

---

[53] Heavy elements with an atomic weight up to that of iron (which has the greatest binding energy per nucleon) form inside stars in their final giant and supergiant phases. As they explode even heavier chemical elements come into existence, including those we use as fuel in nuclear reactors in our power stations.

vaporize (by emitting radiation and so also matter) as an explosion. Nonetheless black holes are made visible by the interstellar matter that glows when it falls into black holes. So-called accretion discs form around them.

> *As far as our own Sun is concerned, astrophysicists anticipate a dazzling future for it. It is expected to shine brighter and brighter until global warming wipes out all life on our planet, that is unless some asteroid or some other disaster involving the entire planet destroys it first. But humankind, which is not among the most resilient animal species, will probably have diedout long before . . .[54] In 3 billion years' time the temperature will have reached some 400 degrees and that heat will remain for another 4 billion years, i.e. about the same period of time as separates us from our planet's creation. After about 7 billion years the Sun's nuclear fuel will be exhausted. The Sun will start to cool and increase in volume. It will swell to 250 times its present diameter and turn into a Red Giant. Due to the loss of solar mass, the orbit of our planet will shift to the distance of Mars and so Earth will elude its fate for another few hundred million years. Finally our planet—no longer green or blue—will evaporate in the red atmosphere of the expiring Sun.*

Most stars came into being about 10 billion years ago, i.e. when the universe was still young. But we can still observe in many parts of the sky the embryonic clouds of hydrogen and the stars emerging from them. The dimensions of the clouds are tenths of parsecs and their mass tends to be approximately that of hundreds of Suns. Although the clouds are of dense matter in terms of the universe, their average density in the initial phases of their evolution can be extremely low

---

[54] Animal species tend to survive for around 10 million years, although the dinosaurs managed to stay around for a bit longer. But there are also some more optimistic scenarios for mankind, such as colonising outer space. We shall see . . .

in terrestrial terms. (A technician would not hesitate to classify such rarefied gas as an ultrahigh vacuum.)

> *An example of an embryonic cloud is the well-known M42 nebula in the constellation of Orion. It is easiest to see in the winter months and a small telescope (or binoculars) suffices. Its red coloration is only apparent on photographs, however. Another beautiful sight is the bluish hydrogen nebulae around the "young" stars in the open star cluster of the Pleiades.*

The Path to the Big Bang

The fact that individual stars are born, develop and expire need not seem too surprising. Similar life cycles can be observed all around us in animate and inanimate nature. What is more surprising is that the universe as a whole goes through a sort of "life cycle". The universe came into existence, develops and, so it seems, will eventually expire.

> *Possibly all mythologies and religions describe the origin of the world. So it must have gratified many believers and confirmed some of them in their faith when "science revived the idea of creation". In 1951 the Catholic Church officially declared that the theory of the "big bang" was in accord with the Bible. Many Muslims view it similarly. By contrast the idea of the creation of the universe perturbed rationalists, as it re-opened old "eternal" question such as how and from what the universe was created, what preceded its creation, how it will come to an end and what will follow it. Scientists are obliged to admit that they have no rational answers to these questions and probably will not have any in the future.*
>
> *From the philosophical point of view, the very concept of "the beginning of the universe" or "creation" seems questionable. Those for whom the idea of a Creator is an obstacle may speak about the "origin" of the universe, but this is not much help because even that concept assumes that something that previously did not exist came into existence. But what was there before the universe? Most cosmologists believe this question is meaningless because there was no "before". That was already the view of St Augustine, who wrote that "God created the world together with time, not in*

> *time", i.e. no "creation", "origin" or even "birth". And not even "a beginning" because that also implies that something "was begotten" (from something).[55] Physicists generally solve the issue by speaking in terms of "singularity". This sounds erudite but it is simply a euphemism for something that cannot be described or explained. (In that respect "singularity" has the rather unfortunate connotations of a "miracle".) So the beginnings of the universe are shrouded in mystery, mysterious not only in terms of physics but also from the point of view of the logic of concepts.*

So let us leave the question of the actual origin of the universe to the philosophers (who frequently leave it to theologians with malicious glee). Let us confine ourselves to what happened afterwards—to the question of the development of the universe—and focus chiefly on the relationship between the evolution of the universe and our dark night paradox.

The Standard Cosmological Model

The overwhelming majority of today's cosmologists base themselves on the so-called standard model, popularly known as the "big bang model". The groundwork for it was laid a century ago, in the final years of the Austro-Hungarian Empire. Albert Einstein, who had already gained a reputation as a physicist thanks to his explanation of the photoelectric effect and above all thanks to his special theory of relativity, spent 1911 and 1912 as a professor at the Charles-Ferdinand University in Prague. From the window of his study he would watch in silence the inmates of an institution for the mentally ill, the gardens of which lay the other side of street behind a wall by St Catherine's church. But his mind was elsewhere, solving the question of the effect of gravitation on the propagation of light. He summarised the outcome of his reflections

---

[55] The question still remains whether the "creation of time" wasn't itself the result of faulty logic.

in a theory that fundamentally altered the physical picture of the world. The theory was published at the end of 1915 and is known as the general theory of relativity. According to this theory there should exist a close relationship between matter (or more precisely between its basic physical manifestation—mass) and space-time curvature. Einstein described that relationship with his well-known gravitational field equations. These are a set of ten non-linear partial differential equations with tensors. They are a hard mathematical nut to crack and they have yet to be solved in general terms. (Solutions to differential equations are not numbers but functions.) When, in 1922, a young Russian mathematician, Alexander Friedman, came up with a special class of solutions for solving these equations (for homogeneous mass density) he came to a surprising conclusion: that the universe should expand or contract. Five years later, the Belgian Catholic priest and theoretical physicist Georges Lemaître (1894-1966) arrived at a similar conclusion (independently of Friedman). These days I expect every schoolchild knows that the universe really is expanding. Einstein rejected the idea of expansion, however, and in order to preserve his preconception—the traditional "scientific" picture of a static universe, he modified his equations by adding a further term—the cosmological constant Λ *(lambda)*. Thus modified, the equations provided a static solution. There was no physical justification for lambda, however, it fell like a *deus ex machina* from somewhere in the metaphysical heights of the physical firmament.

> *There followed from Friedman's solutions two contrasting scenarios for the development of the universe. If the universe was expanding it could expand for ever and the universe would go on thinning out. Alternatively the expansion could stop and give way to contraction ending in a "big crunch". And some physicists believe that the entire development of the universe could be repeated, thereby returning to the ancient concept of a cyclical universe. Cyclical models became popular particularly in the second half of the twentieth century. And in 2001 Neil Turock and Paul Steinhard produced a cyclical model that they called "ekpyrotic" according to Stoic usage. However, today's mainstream*

> *cosmology bases itself on the assumption that what awaits us is not a "big crunch" and possible further cycles, but extensive diffusion in a continuously expanding universe.*[56]

Both Friedman and Lemaître came to their conclusion only "on paper", but their theoretical discovery was soon corroborated in 1929 by observational astronomy. The findings are attributed to Edwin Hubble, who, together with Milton Humason, observed that the spectral lines of light from distant galaxies are shifted to longer wavelengths in dependence on their distances. This phenomenon is known as the "red shift". They were not the first to observe it,[57] but they were the first to explain these shifts as being due to the Doppler Effect caused as galaxies receded. Einstein had no choice but to recognise his error. The universe could not be static and a cosmological constant had no place in his equations.[58]

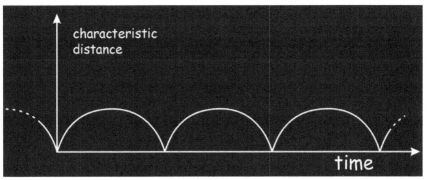

Model of the oscillating universe (cyclical model)

---

[56] See Kirshner.
[57] The displacement of spectral lines towards longer wave lengths (to the red end of the visible spectrum) in radiation from distant galaxies had already been recorded in 1912-1923 by Vesto Slipher at the Lowell observatory in Arizona.
[58] According to George Gamow, Einstein considered the introduction of cosmological constants to have been the most serious error he ever made.

> *The cosmological constant did not disappear for good and all, however. Seventy years later, in 1998, cosmologists dusted it off and gave it opposite sign. Its job now was to describe the acceleration of cosmic expansion, which emerged surprisingly from the latest astronomical measurements. (There is no explanation yet for why the expansion of the universe should accelerate. Astronomers speak of "dark energy" causing the universe to expand. It looks interesting—"energy" is a fashionable term nowadays. Nevertheless it is a label only, not an explanation of anything. So the question of cosmic expansion is back on the agenda. History never repeats itself exactly, however. In Einstein's days the assumption of a static universe was confronted by a theory that would be corroborated by interpretation of astronomical observations. Today there is a theory confronted by observations, or to put it more cautiously, by a certain interpretation of observational findings.*

Hubble's famous discovery was made possible thanks to the biggest telescope of those days, erected on Mount Wilson in California. It had a diameter of 100 inches, or 2.5 metres. The enormous amount of light captured by it not only enabled distant galaxies to be seen, but also allowed their spectra to be recorded and shifts of spectral lines to be measured. The foreign galaxies are so distant that it would take millions of years to record any change in their positions by means of direct comparison of photographs.[59] This means that the only observational proof of cosmic expansion that astronomers have available is the red shift. Is it reliable? And is it also proof that the universe came into existence at some point and will also come to an end at some point? Scepticism about Hubble's explanation and about the entire dynamic model of the cosmos was widespread from the first. These doubts gave rise to a series of alternative explanations that sought to demonstrate that in spite of the (undeniable) red shift, the universe was actually eternal and static, or that it was at least in some kind of steady state (in

---

[59] Remember that Halley only needed data hundreds of years old to record stellar movement.

the same way that the number of a city's inhabitants can be constant, even though people are born, die or move.)

The best-known alternative explanations include the theory that photons age, propounded in 1929 by Fritz Zwicky (1898-1974). That theory did not accept the expansion of the universe and described the red shift (and hence the energy loss of photons) in terms of some kind of "aging" mechanism. At first sight it seems obvious: aging is a universal phenomenon and is usually accompanied by a loss of energy. The computer on which I am writing these words is aging and in a few years will be a write-off, the battery in my watch is aging and eventually the hands will stop moving. I too am aging and my energy is decreasing. Even the Sun is aging and in a few (billion) years it will conk out. There are still some engineers who argue that if all transfers of energy involve a loss why should photons that fly for millions and billions of years be the exception? Nevertheless this argument has a fundamental flaw. As we have demonstrated on several occasions, our experience of the human scale world cannot be simply transferred to the scale of the whole universe. The megacosmos is not an enlargement, and the microcosm is not a reduction, of the macrocosm (as Euclidian geometry tries to convince us[60]). Their physical characteristics are quite different. Neither atomic nuclei nor subatomic particles age. They can collide and disintegrate but they do not change between those events. Moreover, were we to consider the loss of photons' energy to be a sign of aging we would come into conflict with the law of conservation of energy. This is a warning sign for a physicist as the law of the conservation of energy is one of the pillars of physics. If I lose energy, however, it has no significance for physics, because my "energy" is not the right energy in terms of physics: that energy only passes through me. Likewise the energy loss in an electrical power line is not real loss. It is simply the transformation of the electrical energy into heat, which raises the

---

[60] Euclidian geometry is scale-invariant: a triangle has the same properties whether it is the size of an electron or its dimensions are several billion light years. Curved space geometries do not have that property—see the section on Euclid and Riemann in the second part.

temperature of the wires and the surroundings. For a power engineer that heat is unimportant and so it is not used in calculations. It would therefore seem that the theory of photon aging should be ruled out straight away because it contradicts the law of the conservation of energy. From a more general view, however, the problem is, that the law of the conservation of energy (or any other law of physics) cannot be proved with absolute conclusiveness or exactitude, or for all cases, i.e. in a logical sense. As Aristotle himself stressed, all laws in nature have only a limited validity and reliability. They are simply rational assumptions whose validity is highly probable, and which are useful in enabling the construction of a science that stands the test of time. Practice then confirms those assumptions retrospectively. So the point could be made that we will never find any photons of that age in laboratories on Earth, and so we have until now been unable to record the aging of photons. Their energy would have to be lost so slowly as to be simply indiscernible. A cogent counter-argument, however, is that overall the aging of photons does not fit into the structure of physics—it causes more problems than it solves.

> *The hypothesis has also been advanced that photons age because they lose their energy through collisions with the particles of interstellar matter. This would overcome the unfortunate conflict with the law of the conservation of energy. It would give rise to new problems, however, because when photons interact, the direction of their movement should alter. To put it another way, there is no mechanism known to physicists whereby photons could lose energy without directional dispersion. When it disperses, a photon behaves like a bullet hitting an obstacle: it releases some of its energy and flies onwards deflected. Therefore photons should also be deflected and the images of distant galaxies should be blurred. Nothing of the kind is observed, however. Moreover, were we to assume that this effect might explain the mystery of the dark sky—which is the main purpose of our considerations—we would be confronted by the already familiar problem: namely, that the absorbent matter should also be heated up.*

> *Another cogent argument has been advanced in opposition to "photon aging". In 1995 supernovae Type Ia were discovered in distant galaxies; these kind of supernovae are used by astronomers as standard candles for determining distances. We are already aware how their luminosities (brightnesses) are dependent on time (light curves). Astronomers have measured the particular time spans in the light curves of these far distant supernovas and have discovered that they are longer than in the case of closer supernovae. It follows from this that the Doppler interpretation is correct, because the apparent slowing down concerns not only the electromagnetic wave frequencies but also their modulations, i.e. light curves.*

The preferred theory that aimed to salvage the notion of a steady-state universe was to be the continuous creation theory.[61] It was the brainchild of Thomas Gold, Herman Bondi and Fred Hoyle. (It was the same Bondi who attributed the discovery of the dark night sky paradox to Olbers.) It accepts the expansion of the universe as a reality while asserting that the universe is in a steady state: the dilution of matter caused by the expansion is compensated by the constant creation of new matter.

> *There exists a beautiful legend about the origin of the steady-state theory: In 1946 Fred Hoyle, Hermann Bondi and Thomas Gold went together to see the classic horror film Dead of Night, which tells the story of a repeating nightmare. On their return home, Gold suggested producing a cyclical film that one could begin viewing at any point. He then speculated that perhaps the universe might be like this, with no beginning and no end.*[62]

---

[61] It has nothing to do with pseudoscientific creationist theories about the origin of life on Earth. This is to do not with the creation of life but with origin of matter itself.

[62] http://www.joh.cam.ac.uk/library/special_collections/hoyle/exhibition/bondi_and_gold

However, no physical explanation of the process of creation was found. Moreover, creating out of nothing conflicts with the law of the conservation of matter. However, in this case also, the effect would be so tiny that it would be impossible to verify or disprove it in the laboratory.

> *The hypothesis that matter might be created spontaneously out of nothing might seem suspect to critically-minded people. Already in ancient times scholars asserted the principle that it is impossible for something to be created out of nothing. But let us look at it from the opposite angle. Isn't the idea of a Big Bang that suddenly occurred out of nothing just as curious? Moreover, philosophers, or rather philosophically-minded physicists, point out that the "nothing" from which the universe was created (or is being created continuously) is not "truly absolute nothing" but some unknown form of "something", something totally unfamiliar, in other words, "matter" in the broadest philosophical sense.*

It is therefore evident that all the alternative theories solve the problem of cosmic evolution at the cost of introducing new major problems. In order to preserve a static (or at least everlasting) universe, we would have to abandon certain pillars of physics: either the law of the conservation of energy, or the zero rest mass of photons (the theory of aging photons), or the law of the conservation of matter (continuous creation theory). Nowadays these alternative theories are museum pieces only.[63] An increasing amount of evidence has convinced the overall majority of astronomers that the universe truly is expanding and thinning out. The expansion of the universe is quite simply a fact as certain as that the Earth goes round the Sun—in the words of the famous Russian cosmologist Yakov Zeldovich (1914-1987). The expansion of cosmic space causes

---

[63] This does not apply entirely to the theory of the cyclical universe. The theories still have life in them but they are located outside the cosmological mainstream.

galaxies to move away from each other and therefore also from our own Galaxy. This results in the spectral line shift mentioned earlier (photon energy reduction) and an overall reduction in the intensity of radiation coming from remote sources in the universe. The radiation is forced to occupy an ever greater volume of space—not only matter is made thinner in the universe, but also energy. It turns out that the red shift plays an absolutely fundamental role in cosmology as a whole. On account of it, extremely remote objects are less visible and at higher red shift values only invisible long-wave radiation is received from them. And when the distance is even greater no photons can reach our eyes. The remote objects disappear beyond the horizon.[64] This happens when the speed of their progress away from us (in regard to the moment when the light was sent) exceeds the speed of light itself. At that moment the light sent has no chance of reaching us.

> *At first sight the situation seems clear: if we run away from a bullet faster than its speed of flight, it cannot hit us. On the other hand, the situation might look suspect to readers with some knowledge of the special theory of relativity, which is based on the principle that the velocity of the mutual motion of bodies cannot reach (let alone exceed) the speed of light. However, the special theory of relativity only applies locally—at the level of the entire universe we must substitute for it the general theory of relativity, in which the situation is a bit more complicated.*

Of course remote objects also appear less clearly visible for purely geometric reasons—due to perspective. Kepler already proved that the intensity of light decreases with the square of the distance. Hence the red shift amplifies the geometric effect, whereby light intensity decreases faster than with the square of the distance. Doesn't that remind you of something? Halley had already tried to solve the dark sky paradox as being due to the faster decrease in the intensity of

---

[64] Cosmologists introduce more types of horizon. The type of horizon mentioned here is different from the type considered by Mädler a Poe.

light! And with a bit of goodwill we could explain in the same spirit the arguments of the discoverer of the paradox, Thomas Digges![65] So both Halley and Digges were right "in their way". But they failed to take the next, fundamental step—using physics to properly substantiate their argument.

---

[65] Digges actually stated that we do not see the light of distant stars "because of their wonderful distance".

A Paradox that Resisted Solution and the Standard Model

GO OUT AT NIGHTTIME AND NOTE THE DARKNESS OF THE HEAVENS; THEIR DARKNESS PROVES THAT THE UNIVERSE IS EXPANDING!

H. Bondi

Let us return to our dark night sky paradox. What role is played in it by Hubble's discovery of cosmic expansion? And what is the role of the red shift of remote objects? Was the renowned cosmologist Hermann Bondi right? Does the dark night sky truly prove that the universe is expanding? And is this expansion the cause of darkness at night, i.e. also the solution of our paradox? Yes and no. As we have seen already, the dark sky at night could prove all sorts of things. The mere fact that red shift or cosmic expansion could be used to explain the darkness of the night sky does not mean that this explanation is the correct one. We have already seen the fate of other explanations . . . And more philosophically: what does it mean that a particular phenomenon is the explanation of another phenomenon? In my view, a particular phenomenon torn out of a broader theoretical context neither proves nor disproves anything of itself. It all depends on the context, and for the context we must return to the cosmological model as a whole.

What emerges most importantly is that since the universe is expanding it must have been denser before. And it must have been hotter, because, to put it simply, its energy, which, for the decisive part, is thermal in nature, was more concentrated. This has fundamental implications as regards the material nature of what the universe consists of. And here the original Einstein and Friedman model proves insufficient. It was too simplified. According to it the universe consisted of space in which "dust" moved due to the effect of gravitation—a great number of particles, mass points, which interacted together only gravitationally. And that "dust", in its turn, through gravitation influenced space, or more precisely space-time as a whole. (We do not speak about interstellar dust; for cosmologists, "specks of dust" usually stand for galaxies.) The model does not take into account the true nature of that "dust" or any other mutual interaction apart from gravitation; it disregards the electromagnetic interaction of particles, nuclear processes, the creation of new particles and atomic nuclei, as well as the creation and evolution of life. The neglect of chemistry is not such a great misfortune. Chemistry does not exist at high temperatures, because chemical bonds are not strong enough to hold molecules together. And this is doubly true of biology. Moreover, life is important only for us living souls, and not for the universe as such. At high temperatures only nuclear physics plays a significant role. Einstein's concept was therefore simply the first approximation, on which to build a proper physical model later. A crucial step towards building such a model was the linking of the physics of the megacosm (cosmology) with the physics of the microcosm, nuclear physics, which engendered the standard cosmological model, whose latest, almost universally accepted variant is The Concordance Model, as it is known. Concordance here concerns views on the individual phases of cosmic evolution and also the values of fundamental cosmological parameters.

According to the standard model, the universe was initially very hot and very dense. It consisted of incandescent plasma—a mixture of sub-atomic particles, which interacted not only gravitationally "at a distance", but chiefly "at close range" through intense nuclear and electromagnetic effects. As the universe expanded the density of the plasma and its energy diminished. And since temperature depends on energy, which also decreased, the universe cooled. After about 380,000 years the temperature had fallen so much that positively-charged particles (protons) bonded with negatively-charged particles (electrons) in stable structures. This gave rise to electrically neutral atoms. Initially they were hydrogen atoms, the simplest of atoms, consisting of a proton orbited by an electron. (Helium and lithium also came into existence to a lesser extent.) Physicists call this process recombination, although the name is not particularly apt in this case, as the electrons and protons had never combined before. The newly born neutral atoms scarcely reacted with light or other electromagnetic radiation any longer. Radiation was "freed" from matter and travelled almost unimpeded through cosmic space. The universe became transparent. The space was filled with an intense glow—the last photons released from the germinal plasma during the period of recombination, at the time of the last dispersal. The energy of these photons corresponded to a temperature of about 4,000 K (slightly higher than the temperature of a light-bulb filament, and slightly lower than the temperature on the surface of the Sun). Stars did not exist yet and the universe went on expanding. The energy of the hydrogen atoms and photons continued to diminish—the universe continued to cool down. This is because the photons emitted during the last scattering arrived from increasingly remote areas. The important fact is that the pace of cooling of radiation is different from the pace of cooling of particles. As a result, the universe remains in a state of thermal (and hence also thermodynamic) nonequilibrium. If we had been alive at that time the sky above us would have been one solid glow without stars. It would have been the image of that germinal plasma, which no longer existed, but would have been visible because of the delay with which light reached our eyes. At first we would have been dazzled by clear white light, then the glow would have become yellower, redder and still darker before being extinguished altogether. At that moment the sky would have emitted only infrared radiation—heat. But the heat would also have diminished in time.

As we said earlier, on a cosmic scale, gaseous hydrogen is unstable. The decreasing kinetic energy of its atoms and molecules could no longer defy gravity and the gas started to cluster in clouds. The clouds went on being compressed until they started to form stars. The dark "age of gloom" came to an end and the stellar era of the universe began. Although the universe was full of myriads of stars it continued to be largely empty transparent space. Vacuum and darkness prevailed. Within that emptiness there travelled photons from the stars along with "old photons" that had been released from the germinal plasma at the time when it was becoming transparent. In the meantime their energy has diminished so much that it is now at a temperature of less than three degrees above absolute zero[66] and their corresponding wavelength is in the millimetre waveband.

The above is a rough outline of the theory whose foundations were laid by the American physicists George Gamow, Ralph Alpher and Robert Herman, which is popularly known as the Big Bang Theory. As stated previously, for a long time this theory was confronted by preconceived notions of the invariability and eternity of the cosmos, i.e. the theories of a steady state universe. The turning point that switched the score in favour of Big Bang Theory was precisely the discovery of these "old photons". This radiation was named cosmic microwave background radiation (CMBR or CMB) or relic radiation—because it was a relic of much earlier times, from the period when the glowing germinal plasma came to an end. The radiation reaches us almost uniformly from all directions and constitutes the microwave background to the entire cosmic scenery.

> *George Gamow predicted that by now the background radiation should correspond to a temperature of an emitter of around 6 degrees above absolute zero.[67] Paradoxically that radiation was discovered by chance. In the mid-1960s two radio engineers at the Bell Laboratories in New Jersey, Arno Penzias and Robert Wilson, captured a very weak electromagnetic signal of unknown origin.[68] This was achieved by using a radio antenna with an area of 25 m², which was intended to detect signals bouncing off the Echo satellite.[69] Penzias and Wilson discovered that the mysterious radiation came uniformly from all parts of the*

---

[66] Meaning 0 K: "zero kelvins" (see the section on Kelvin in the second part)
[67] It should have the nature of the radiation of an absolutely black body at 6 K. Every body with non zero absolute temperature emits electromagnetic radiation. Its intensity and spectral composition is dependent on the temperature of the body and it is also influenced by the nature of the emitter surface. An absolutely black body is the best emitter and the best absorber.

> sky and its wave length corresponded to the thermal radiation of an absolutely black body at a temperature of about 3 K.[70] Meanwhile, at nearby Princeton University, a research team led by Robert Dicke and including Jim Peebles, Peter Roll, and David Wilkinson) independently calculated the anticipated temperature of the background radiation and started to design equipment to detect it. The team's members heard about the mysterious radio noise measured by the Bell laboratory employees and proposed the predicted background radiation as an explanation.

So, what is the recent solution of dark night sky paradox? Although there are many stars shining in the sky there are not enough of them to fill the entire sky. Because of their size, the stars are very remote from each other, so remote, in fact, that the visual range in that "sparse forest of stars" should be enormous—about 10^{23} light years in terms of distance, i.e. also 10^{23} years into the past. But the universe is much younger, stars shine only for about 10^{10} years, so we must be seeing beyond the stars, or into a period before the stars came into existence. What we see there is darkness, however. Thanks to it we have night, we have acceptable temperatures on Earth, and we are alive. That darkness is not a view of "nothing", however, no empty space. It is an image of the long extinguished fire of the germinal plasma, from the period when light freed itself from the bondage of matter. That fire now appears black, because its radiation is so diminished thanks to the expansion of the universe. The glow has cooled, but it can still be observed with sensitive equipment in the range of microwaves. These observations indicate that the entire sky glows feebly and almost uniformly. The temperature corresponding to that invisible glow is less than three degrees above absolute zero. It is

---

[68] published in 1965
[69] That satellite was the first attempt at a telecommunications satellite. It consisted of a balloon of metalized plastic foil, 30 m in diameter. Once in orbit it was inflated and signals were reflected passively from its surface.
[70] Latest measurements give the temperature as 2.725 K.

no longer the glow of an incandescent inferno, but the pallid glimmer of bitter frost.

Return of the Aristotelian Universe?

When we look into the distance we also look into the past. Therefore the universe as we are able to observe it, and the way it affects us and exists for us, is enclosed in a hollow ball of germinal plasma. Doesn't that picture remind you of something? Isn't this the return of the Aristotelian model after a thousand years? The ultimate sphere is not a sphere of fixed stars this time, however, but the sphere of the "last scattering", a sphere of plasma at the end of the Big Bang. The similarity cannot be denied.

We long ago abandoned the Aristotelian model because of its geocentrism—i.e. its glaring egocentrism. After all, we are not the centre of the universe! However, as we can see, it has come back to us through the back door in a new guise, including that irksome singularity. In Aristotle's case it was purely spatial (the end of space—the last sphere, the heavens), but now encompassing both space and time (the sphere of the last scattering). How come? Are we guilty of some methodological transgression? After all, the starting point of any scientific research is that its results should be independent of the observer, of the reasoner, or of other subject. Only then is it objective and universally valid. That step was essential for creating science but it was a requirement that eliminated ourselves from the game. However, all knowledge has phenomena as its basis and phenomena appear to some observer, a subject. We are the ones, here and now, that exist, observe and reason, we are the ones that see the dark sky at night, we are the ones that reside in the centre of the sphere of the visible universe. There is no way that we can consistently eliminate ourselves from the game . . .

TO BE IS TO BE PERCEIVED

George Barkeley[71]

> *However, the present model of the universe really has even more similarities with the old Aristotelian one. In the Aristotelian view, a freely moving body, i.e. a body without forces acting on it, stopped its motion. And the situation is to some extent similar to the situation in the expanding universe. Present-day cosmologists work with a so-called co-moving coordinate system which is expanding together with the world of galaxies. A freely moving body in such a system also stops its motion. It can reach only an area which is receding at the same speed as the original speed of the body.*

---

[71] George Barkeley (1685–1753) was Anglo-Irish philosopher, protagonist of "immaterialism" (called "subjective idealism" by others). In the field of physics he argued against Newton's doctrine of absolute space and time. His arguments were a precursor to the views of Albert Einstein.

Darkness at Night and its Total Cause

So why is it dark at night, why does darkness prevail in the universe? In literature you can find many answers, mostly particular phenomena taken out of context, whether it is "aging photons" (an unproven phenomenon), expansion of the universe, the Doppler Effect, lack of energy or something else. However, the best explanation, or the total cause, if you like, is the whole structure and history of the cosmos.[72] And present-day cosmology is now capable of describing its basic features.

So is the dark night sky paradox solved after four hundred years? I think so. But that doesn't mean there are no surprises in store for us in that sphere.

EVERYTHING MUST BE MADE AS SIMPLE AS POSSIBLE. BUT NOT SIMPLER.

Albert Einstein

---

[72] "Total cause" or "causa totalis" is the cause, which itself is responsible for the effect. The concept was introduced in 13th century by the Scots philosopher John Duns Scotus (1266–1308). Scotus is one of the most important and influential philosopher-theologians of the High Middle Ages.

The Players in our Story

The second part of the book will return to the familiar and less-familiar players in the story of attempts to explain why the sky is dark at night. It also includes a brief description of some of the physical concepts and principles not dealt with in the previous section.

Archytas of Tarentum
(C. 400-365 BCE)

The distinguished Pythagorean Archytas lived about one hundred years after Pythagoras. He was a popular ruler of Tarentum in southern Italy, and was known as a strategist, a scholar and a skilled technician. Like the other Pythagoreans, he also had an interest in astronomy, mathematics, mechanics and acoustics. He was a friend of Plato, whom he acquainted with the mysteries of the Pythagorean sciences. And when Plato was jailed in Syracuse, he interceded with a letter and helped obtain his release.

From our point of view, Archytas was of particular importance because he drew attention to the paradox that emerges from the concept of a finite universe. He argued that if the universe had a frontier, there would have to be something beyond it. His argument is still valid

Aristarchus of Samos
(3rd century BCE)

Aristarchus was a Greek scholar from Samos, the island where both Pythagoras and Epicurus were born. He taught that the universe was infinite and that the Earth revolved around the Sun and not the contrary. He did not manage to bring his compatriots round to the idea of heliocentrism. Indeed they seem to have been outraged by the idea. Three generations later Aristarchus' idea was adopted by the neo-Babylonian mathematician Seleucus (150 BCE), but he remained a lone voice. Seventeen centuries would pass before Nicholas Copernicus breathed fresh life into Aristarchus' concept.

Aristarchus was aware one of the interesting implications of his presumed heliocentrism: if our planet revolves around the Sun, its motion should be reflected in an apparent motion of the stars in the sky. This phenomenon is now known as stellar parallax (annual heliocentric parallax).

> *Imagine we are travelling by train. We are seated by the window observing the landscape. The train's motion appears to us as a backwards movement of the surrounding objects. The poles alongside the track move backwards more quickly than the houses that are further away and the hills on the horizon don't seem to move at all. The same should apply to the stars. Due to the motion of our planet, the nearer stars should move backwards relative to the Earth's motion. The motion of the remoter stars should be slower and those very far away should not seem to move at all. And since our planet revolves around the Sun, the stars should also describe circles or ellipses (an ellipse is a projection of a circle). The parallax (parallax size π) is known as the angular radius of the circle or the longer half-axis of the described ellipse.*

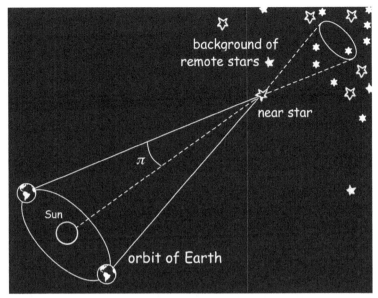

Stellar parallax

Despite all his efforts Aristarchus failed to measure any parallax. Other astronomers were equally unsuccessful, so the absence of parallaxes became the most cogent argument against heliocentrism. It was a rational and astronomic argument not just a psychological or theological one. It took two millennia before parallaxes were proved to exist after all. They are so small, however, that they cannot be measured using a simple astronomical instrument—the stars are simply too remote. It is striking that Aristarchus had already come up with the explanation.

> *Stellar parallaxes were measured in 1838 by the German astronomer and mathematician Friedrich Bessel (1784-1846). He made his measurements using a telescope, of course. From the measured values he calculated that one of the stars in the Cygnus constellation was 600,000 times further away from the Earth than the Sun.*

> *The measurement of parallaxes is still used as the most direct way of determining the distances of the nearest stars. This was the basis for defining the astronomic unit of distance, the parsec. One parsec corresponds to the distance at which the mean radius of the earth's orbit subtends an angle of one second of arc. And conversely: stars at that distance have a parallax of one second (hence par-sec: parallax of a second). One parsec represents about three light years (ly). The nearest stars are at a distance of several parsecs.*

Aristotle of Stageira
(384-322 BCE)

Aristotle, the most renowned scholar of the ancient world, contributed to almost all areas of human knowledge. In the field of cosmology he perfected and supported with his reputation the geocentric model of the universe.

He was born in Stageira, Chalcidice in northern Greece. He then studied and taught at Plato's famed Academy in Athens. Following Plato's death he spent three years at Assos (Asia Minor, today Turkey) before accepting an invitation from Philip II to go to Macedonia, where he became tutor to Prince Alexander in the capital city of Pella. Alexander assumed the reins of power when his father was assassinated and united Greece under Macedonian rule before setting out to conquer the whole world. Aristotle returned to Athens, which was already under Macedonian sway. With the support of the Macedonian governor, he founded his own school, known as the Peripatetic school, or the Lyceum. He was very active as a scholar and teacher. That fruitful period of activity was eventually interrupted by the anti-Macedonian uprising that broke out after Alexander's sudden death. Aristotle paid dearly for his friendly connections with the Macedonian court. The Athenians viewed him as a collaborator and threatened to put him on trial. Aristotle had already prepared his defence speech, but recalling that Socrates had also been convicted of impiety, he decided "not to give the Athenians a second opportunity to sin against philosophy" and set sail for the nearby island of Euboea, where he died two years later

Aristotle imagined the universe to be spherically symmetric and finite. He placed the spherical Earth at the centre and had the Moon, the Sun and the other planets revolving around it. Everything was enclosed within the sphere of the fixed stars. Aristotle in his physics did not introduce concept of inertia, and so ascribed all motion to the effect of forces.

> *The need for forces to cause motion is the fundamental difference between Aristotelian and Newtonian physics. In Newtonian physics, no force is needed for straight uniform motion, it takes place thanks to inertia. (However, this only applies if the Greek expression "dynamis" is translated as "force" in the physical sense of the word. But "dynamis" has many other meanings, such as potency, capacity, ability, power, strength, possibility . . .)*

In the framework of Aristotelian mechanics the heavenly spheres were unable to turn of their own accord, but needed something to turn them. Aristotle assigned this role to the "Prime Mover". And that otherwise commonsensical philosopher even ascribed to that strange instigator the power of thought and divine attributes. That was not much help to the Mover: it never really became a god in the eyes of the public. In later times the Christians also had a problem with accepting it. Nevertheless Aristotle's ideas continued to spread. This was particularly thanks to Arab scholars, who translated his works and provided explanations and commentaries. Eventually the church changed its tactics. Aristotle's teachings were no longer banned but merely "corrected", that important task being assigned to Thomas Aquinas (1225-1274), who purged Aristotle from "the accretions of pagan commentators" and paraphrased him in a Christian spirit. He turned the Prime Mover into God. (This interpretation of the Prime Mover had already been adopted by the Spanish Jewish philosopher Maimonides (1135-1204).) Thus Aristotle became a prominent Christian philosopher and Thomas achieved sainthood. Along with Aristotle the Christians also adopted Aristotelian rationalism and the geocentric universe. However it became increasingly clear that this was not such a great prize after all. The church defended out-dated Aristotelianism and geocentrism so vehemently that it ended up undermining its own authority. Thus two thousand years after his death the philosopher became an impediment to knowledge.

Aurelius Augustinus, St Augustine
(354-430 CE)

Until the adoption of Aristotelianism in the 13th century, the Roman bishop Augustine was the major authority of western Christianity. His writings were not confined to theology but also include profound philosophical speculation. Augustine's ideas about time are still relevant in terms of current cosmology.

Augustine came from Thagaste in present-day Algeria. He mis-spent his student years with hard drinking, rowdy behaviour and enjoying himself with women of ill-repute. In the depths of his unconscious, however, a conflict was being played out that resulted in an unexpected volte face. While out walking he heard a child's voice saying: "Take up and read". He took it to be a signal from above and he picked up a book and started to read it. It was probably not by chance that the passage on which his eyes first fell was St Paul's epistle to the Romans. Augustine immediately understood the direction of which this "divine touch" was steering him. He abandoned his dissolute life and he set off on a path of devoted work for the church. He started to write theological and philosophical texts. Shortly afterwards he was ordained bishop of Hippo, and in this position he was both admired and loathed. (Augustine's views on predestination proved particularly controversial, as well as his views on the church's supremacy.) His life ended as the Western Roman Empire was coming to an end: he died in Hippo on August 28th of the year 430 at a moment when the town was under siege from barbarian marauders.

In his youth he won his battle with temptation, but he never managed to solve his inner conflict between theologian and philosopher and he strove in vain to reconcile Christianity with Greek philosophy.

Augustine aspired to establish an objective approach to learning. But he gave up that quest and turned his attention inwards. Only within himself would he find God and the fundamental certainty. Augustine's views about time are still remarkable. But he refused to formulate a definition of it:

WHAT, THEN, IS TIME? IF NO ONE ASKS ME, I KNOW WHAT IT IS. IF I WISH TO EXPLAIN IT TO HIM WHO ASKS, I DO NOT KNOW.

That well-known statement not only illustrates how mysterious the very notion of "time" is, but also how impossible it is to define basic concepts. Augustine was clearly aware that the existence of time depends on the existence of the world in which time somehow manifests itself. And he also speculated on the beginnings of time. His statement:

GOD CREATED THE UNIVERSE NOT IN TIME, BUT TOGETHER WITH TIME.

is still taken seriously by most cosmologists. (This opinion had already been voiced in Plato's dialogue *Timaeus and Critias*.) It is also valid, of course, in the absence of the Lord God: the universe quite simply came into existence together with time. So there is no point in asking why God didn't create the heavens and earth earlier or later, or what He was doing before He created them. (Although some catechists apparently know the answer: "He created Hell for those who ask such nit-picking questions.") There was simply no "before". In that way Augustine consistently abandoned antiquity's cyclical concept of time and asserts the notion of development: History started with the Creation and will come to an end when Christ returns to Earth.

Hermann Bondi
(1919-2005)

Bondi developed the steady-state theory of the universe along with Fred Hoyle and Thomas Gold, and he also made an important contribution to the general theory of relativity. It was he who attributed the discovery of the dark night sky paradox to Olbers—wrongly, as we now know.

He was born in Australia but moved to Austria with his parents as a child. When Hitler came to power, Bondi's family, of Jewish origin, fled to Britain and the young Hermann entered Cambridge University. At the start of World War II Bondi was interned as an "enemy alien"; although he had fled from Hitler he had retained his Austrian citizenship. But it's an ill wind that blows nobody any good and in the internment camp he made friends with Thomas Gold, who was in a similar predicament. They were released in 1941 and then worked together on the development of radar. After the war Bondi taught mathematics at university while working on the steady-state theory of the universe. That theory remained popular until the end of the 1960s, when it was dealt a mortal blow by the discovery of cosmic microwave background radiation.

Tycho Brahe
(1546-1601)

Born Tyge Brahe, he was the most famous of the Danish astronomers. He also has a Czech connection, however, because he spent the final years of his prolific life in the Prague of Emperor Rudolf II, the city where he is buried.

Tycho's career was preordained by an eclipse of the Sun that he observed as a little boy. He was to experience a much rarer astronomical event in November 1572, when a supernova exploded in the constellation of Cassiopeia. It shone so brightly that it was visible even in the daytime sky.

He studied in Copenhagen and Leipzig. After receiving his inheritance he devoted himself solely to his hobbies, astronomy and alchemy. He travelled around Europe and only returned to Denmark when King Frederick II made him a proposition and had built for him an observatory named Uranienborg (the castle of Urania, the muse of astronomy) and a "star castle"—Stjerneborg on the island of Hven. (A small island 4.5 km long by 2.4 km wide. It is now called Ven, and belongs to Sweden.) Brahe installed his observation equipment there very unusually below ground level in order to insulate them from the wind and from temperature variations. Brahe had his own farm at his disposal on the island, as well as a windmill, church, fishpond, library, alchemical laboratory, engineering workshop, printing press and paper mill. Many of the serfs worked for him and selected young astronomers assisted him in specialised tasks. Brahe abandoned the island in 1597 when a new king ascended the throne and he no longer enjoyed royal patronage. The famous observatory suffered a sad fate. The villagers had regarded Tycho as a despot so when he left they demolished the observatory.

The year 1599 marked the beginning of the last phase of Tycho's life. Emperor Rudolf II invited him to Prague to serve as imperial astronomer and astrologer. A year later Johannes Kepler also arrived in Prague and assisted Tycho in evaluating his measurements. Tycho's main aim was to measure the parallaxes of heavenly bodies, since the existence of parallax would confirm that the Earth revolved around the Sun and would enable the distances of heavenly bodies to be measured. Despite all his efforts, however, Tycho was unable to measure any parallax. For that reason he did not accept heliocentricity. But by then Aristotelian/Ptolemaic geocentrism was no longer acceptable. So Tycho created a compromise model: the Earth was the centre of the universe and the Sun and Moon revolved around it. The other planets revolved around the Sun.

Tycho's life came to an end in the autumn of 1601. He was not yet 55 when he suddenly died. Various legends surround Tycho's death. According to one of them, his bladder burst, while others assert he was poisoned by mercury. However, these stories are implausible. He was buried in the church of Our Lady before Týn on the Old Town

Square in Prague. (As a protestant, he could not be buried in the main city's cathedral of St Vitus at Prague Castle.)

Tycho's name was given not only to a lunar crater but also to asteroid No. 1940 RO and to a catalogue of the Hipparcos satellite that measures stellar parallaxes.

Carl Vilhelm Ludvig Charlier
(1862-1934)

Carl Charlier was one of the most important Swedish astronomers. After graduating from the University of Uppsala he worked in Stockholm, and was director of the observatory of Lund University for thirty years.

His work focused on celestial mechanics, photometry and practical optics. Later in life he also developed an interest in statistics, particularly the theory of errors. He made extensive studies of the distribution and motion of stars in the solar neighbourhood. He posited a hierarchical model of the universe, according to which it was composed of bigger and bigger entities (a structure now described as fractal). He also proposed that model as a solution of the problem of darkness at night.

Charlier was buried in Lund. A lunar crater is named after him, as well as a crater on Mars and asteroid 8677.

Jean Phillip Loys de Chéseaux
(1718-1751)

The astronomer Jean Phillip Chéseaux worked in the Swiss city of Lausanne. Among the subjects of his research was the dark night sky paradox. He calculated that if the universe were homogenous and infinite the Earth would be bathed in a glow 91,850 times more intense than on a sunny day.

In 1746 he presented to the French Academy a list of nebulae, seven of which were his own discoveries. He also discovered two comets. He is also known for his chronological analysis of the Bible.
He died at the early age of 33.

Nicolaus Copernicus
(1473-1543)

The father of heliocentrism, the Polish astronomer Nicolaus Copernicus (also Mikołaj, Mikoláš, Mikolaj, Nicholas, Nikolas Kopernik, Koppernigk), shifted the centre of the universe from the Earth to the Sun, set the Earth rotating and halted the movement of the sphere of fixed stars.

He was born in Torun in the north of Poland. He attended the University of Krakow but did not complete his studies. He later studied the humanities, medicine and law in Italy. In 1501 he returned to Poland, becoming a canon in Frombork, a small town in the north, where he spent the rest of his life. He lived detached from the world of science, fulfilling his ecclesiastical vocation, practising medicine, and writing papers on currency reform. But his greatest source of enjoyment was observation of the heavens. From his observatory in the tower of the cathedral in Frombork he observed the motions of celestial bodies, noting and processing his observations. He outlined his ideas about the heliocentric hypothesis only in a modest manuscript, the *Little Commentary* (*Commentariolus*). He also hung back from publishing his supreme achievement, *On the Revolutions of the Heavenly Spheres* (*De revolutionibus orbium coelestium*). Was he afraid of the Church's reaction or did he regard his work as incomplete? Copernicus' sight deteriorated and his last observation was the solar eclipse of 1541. In May 1543 he suffered a stroke. An advance copy of his masterpiece is said to have been brought to him on his deathbed. It is unlikely that he was aware of it. But by his death he escaped persecution. He was buried in the cathedral in Frombork.

> *Copernicus' manuscript "On the Revolutions of the Heavenly Spheres" was the subject of heated controversy. It yielded its author posthumous fame and execration. The very fate of the manuscript is fascinating. It was bought in Heidelberg by the Czech student John Amos Comenius,[73] who studied it but was not convinced by it on the grounds of scriptural authority. That manusscript ended up in the collections of the Nostitz Library in Prague. In 1953 it was loaned for an exhibition in Poland, where it has since remained (the Poles sending in its place a manuscript of Comenius' "Labyrinth of the World and Paradise of the Heart"). Copernicus' manuscript now resides in the library of the Jagellonian University in Krakow, where the author was a student.*

Copernicus turned the traditional concept of the universe on its head. He placed the Sun where the Earth had previously been. He declared that our planet rotated on its axis and revolved around the Sun. However, he also decided that the sphere of fixed stars was motionless, and he also shared Aristotle's view that there was nothing beyond the fixed stars.

Thus our blue planet emerged victorious from the ancient controversy over what was in motion—the heavens or the Earth. The latter revolved "in reality", the heavens only "apparently". The situation is actually a bit more complicated and therefore also more interesting. The inappropriateness of the term "apparent" (motion) is illustrated by an argument that I read recently on some astronomical website. A layperson expressed surprise that when making observations with an astronomical telescope it was necessary to compensate for the motion of the celestial sphere by

---

[73] Johannes Comenius, in Czech Jan Amos Komenský (1592-1670) was a Czech teacher, educator and writer. A protestant refugee from re-catholicised Bohemia, he was one of the earliest champions of universal education. He lived and worked in many countries in Europe, including the Netherlands and England.

rotating the telescope around the polar axis. "After all, that motion is only apparent, unreal, so why must one make a real movement to compensate for it?" So let us concede that the movement of the sky is not "simply apparent", that it is relative to the Earth. That deals with one naïve objection, but it opens up scope for a much weightier argument: If all motion were relative and dependent on a frame of reference it would be immaterial whether our planet revolved or the heaven or even both. So how about this relativity of motion?

> *Let us take the motion of a train as an illustration. If our coach has good springs it is hard to tell whether it is in motion or stationary, and when we look out of the window we may happily assume that what is moving is the landscape around. The fact that uniform linear motion is relative was noticed by many of the scholars of the past. Galileo describes it graphically in his dialogues. Serious difficulties arise when it comes to accelerated or decelerated non-uniform, i.e. curvilinear, particularly rotational, motion. Even the best suspension won't help in a bend, or when accelerating or braking. The absolute (non-relative) nature of accelerated motion was long used as an argument for the existence of absolute space, in respect of which motion takes place. But this was a problematic concept, because from the point of view of classical physics space was "nothing". (Space, in Newton's view, was "Sensorium Dei", God's sensory organ, as it were. In other words, not physical in the usual sense of the word, nothing that could be incorporated into physics.) It was not until Einstein's general theory of relativity that "nothing" was replaced by a tensor gravitational field. This formulated the laws of physics so that they could also apply in coordinate systems in non-uniform motion (i.e. in rotating systems as well). So from the perspective of general relativity it really is immaterial whether the Earth rotates or the rest of the Universe. From a practical point of view, however, it makes sense to stick where possible to so-called inertial coordinate systems, where the law of inertia applies, along with the other mechanical laws we are familiar with from our*

> schooldays. Otherwise the description of physical events would be very complicated, if not impossible. Ease of description is therefore essentially the only, albeit cogent, argument in favour of a system in which our planet rotates while the surrounding universe is "stationary".

Heliocentrism did not have an easy ride at the outset. It had to contend with substantial astronomical arguments, particularly with the fact that no parallax had been observed. In Copernicus' day, however the most compelling argument was its incompatibility with *Holy Writ*, and so his views were rejected by Catholics and Protestants alike. (The Catholic church did not formally accept heliocentrism until 1846.) Neither Comenius nor the great reformer Martin Luther himself had any sympathy for Copernicus:

"THE FOOL WANTS TO TURN THE WHOLE ART OF ASTRONOMY UPSIDE-DOWN! BUT SACRED SCRIPTURE TELLS US THAT JOSHUA COMMANDED THE SUN TO STAND STILL, AND NOT THE EARTH!"

Copernicus' name is borne by one of the most beautiful craters on the Moon, and also by a satellite detecting the ultraviolet stellar spectra (operational between 1972 and 1981).

Nicholas of Cusa
(1401-1464)

The last great mediaeval philosopher, Nicholas of Cusa (Cusanus), was a German cardinal, diplomat, theologian, all-round scholar and inventor (He is said to have invented spectacles for the shortsighted.)

Born in the German town of Kues, he studied law in Heidelberg and Padua, although he was more interested in theology, philosophy and the natural sciences. After losing his first case on his return to Germany he hereafter devoted himself entirely to philosophy and service to the church. He was one of the first to attempt to reconcile the various monotheistic religions. He took part in a mission by a papal delegation to Constantinople, then under threat from the Turks, to discuss reconciling relations between the western and eastern churches. He worked for reformation of the church in Germany. Because of his principled stands he became such an irritant that his opponents tried to poison him. He died in the Italian town of Todi and is buried in Rome at St Peter ad Vincula, famed for Michelangelo's statue of Moses.

Cusanus' ideas are a rich mixture of profound philosophical insights, religious prejudice and fantastic notions. He understood the Universe to be "the unfolding of God" and a mediator between God and man. Mathematics plays a major role in it because God created the world according to precise mathematical laws. The Earth has no privileged status as the Universe has no boundaries and no centre. The Earth travels in an orbit like the other celestial bodies. To what extent Cusanus' cosmology inspired Copernicus and Kepler is still a matter of controversy.

Some of his views belong to the realm of fantasy. He believed that there were people living on the Moon, the Sun and other stars, and that they were more perfect than we terrestrials.

He held that nothing in the world is perfect or precise and therefore human knowledge is necessarily limited; Cusanus used the term "learned ignorance" ("docta ignorantia") to describe it. This was the actual aim of learning. In unlearned ignorance, the less we know, the more confident we are in our knowledge. In learned ignorance, the more we know, the less confident we are in the truth of our knowledge.

> Noteworthy are Cusanus' ideas contained in his dialogue "The Layman on Experiments with Weight-Scales" (Idiota de staticis experimentis) in which the author poses as a layman (idiota), a word that was not at all derogatory, but meant someone unencumbered by the prejudice of education who asks irrational questions and is not afraid to voice unusual judgments.
>
> "For although I admit that I am an uneducated layman, I am not at all afraid to state my view. Well-educated philosophers and those who have a reputation for knowledge rightly deliberate more cautiously, fearing to fail."
>
> Cusanus was undoubtedly correct in that respect. However, the role of laypeople in science has a downside too. Insufficient specialised knowledge can render the non-specialist approach naïve. Moreover, an "idiota" is often encumbered with other preconceptions, many of them pseudo-scientific. Nevertheless, as the cosmologist Steven Weinberg (b. 1933) has stated:
>
> ... THE GREAT THING IS NOT TO BE FREE OF THEORETICAL PREJUDICES, BUT TO HAVE THE RIGHT THEORETICAL PREJUDICES.
>
> Cusanus further developed ideas about how all differences between materials must somehow be reflected in their weight (density). Careful weighing should therefore enable one to distinguish not only between different materials, but also between good and bad water, for instance, or between the urine of healthy person and of an ill person, etc. Cusanus might therefore be regarded, at the risk of slight exaggeration, as the discoverer of the importance of measurement for science.

Democritus of Abdera
(c. 460-362 BCE)

Democritus is regarded along with Leucippus as one of the founders of atomism in Antiquity. Only fragments of his work have survived, although he was one of the greatest thinkers. This is due particularly to the negative reputation of his "materialist" teachings.

Ancient atomism was founded on a speculative approach to the world. However, these speculations were often heading in the right direction, and reflected profound insights. Everything material (which also included souls and the gods) was thought to consist of indivisible particles—atoms. Things with various perceivable properties were created by the combination of those imperceptible atoms, in a similar way to the creation of comedies or tragedies by a combination of letters. Atoms were governed by physical laws. (Mathematical formulations of physical laws were unknown at that time of course.) Therefore their movement and hence future events were precisely predetermined; this is known as (absolute) determinism.

The atomists anticipated multiple "worlds", some of which were inhabited. They also envisaged an evolving universe with stars that are born and die. Even the Milky Way was composed of "atoms" according to Democritus—tiny stars that were joined together "as if many fine grains of salt had been poured out in one place".

The old atomists inspired the Epicureans, who in turn inspired Newton at the time of the Renaissance. Atomism was the basis for virtually all modern chemistry and physics.

Thomas Digges
(1546-1595)

The British mathematician and astronomer Thomas Digges was born in the same year as Tycho Brahe. His father died shortly afterwards and the boy grew up under the guardianship of the mathematician, natural scientist and well-known occultist, John Dee. Fortunately, the young Thomas was more interested in science than in the occult. He was to become the first promoter of heliocentrism in England. Digges made significant improvements on Copernicus' system, getting rid of the sphere of fixed stars and scattering the stars throughout infinite space. It was the fact that he planted the heliocentric system into infinite space full of stars that led him to postulate the dark night sky paradox.

Digges also led a political and military career. He became a Member of Parliament and was head of supplies with the British forces in the war with the Spanish Netherlands.

Sir Thomas Digges died in 1595 at the age of 49. He was buried in the Church of St Mary Aldermanbury, in the City of London.[74]

Johann Christian Andreas Doppler
(1803-1853)

The Austrian mathematician and physicist Christian Doppler was born in Salzburg. He spent the most productive period of his life in Prague, where he made his most important discovery, the well-known Doppler Effect.

Like Socrates, Christian Doppler was the son of a stonemason. And like Socrates, he opted for a different career. He was physically weak and so he chose to study astronomy and mathematics at Vienna University. He started working professionally in Prague in 1835, teaching first at a second-level building school and later at the Prague Polytechnic, the forerunner of the Czech Technical University. In 1840 he was elected an associate member of the Royal Bohemian Society of Sciences and a year later he was appointed to a professorship. During his time in Prague he published a number of scientific papers. In May 1842 it was in his "stellar" essay *On the Coloured Light of the Double Stars and Certain Other Stars of the Heavens*, that he first described and explained the phenomenon now associated with his name.

---

[74] Digges got no peace after his death. St Mary's church was destroyed by the Great Fire of London in 1666. In 1677 it was rebuilt by Christopher Wren. During the Second World War, it was hit by a German incendiary bomb and was gutted, leaving only the walls standing. In 1966 these stones were transported to Fulton, Missouri, and rebuilt in the grounds of Westminster College as a memorial to Winston Churchill.

Why is it dark at night?

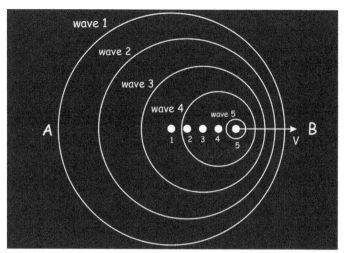

Doppler Effect. Source of waves moves from point 1 to point 5. Receiver of waves B which is in direction of the motion (arrow *v*) will record shorter wave length (higher frequency), the receiver in the opposite direction A, a longer wave length (lower frequency).

> *When the source of waves and their recipient move towards each other, there is a shift in the frequency of the wave received. When we listen to the horn of an approaching car we hear a high tone and the tone gets lower as it passes us and begins to move away. This effect concerns not only sound, but also light and other kinds of waves. The Doppler Effect is caused by "the stretching or compression of waves" between the wave source and the recipient. If the wave source is moving forward the waves in front of it become denser and strike the recipient with a higher frequency (higher tone). The opposite is true as it moves away.*

In his famous work Doppler asserted that stars that have a bluish light are moving towards us, while those with a reddish light are moving away. Subsequently, however, it was shown that this explanation was incorrect. The colour of stars actually depends much more on their surface temperature, and the Doppler Effect is only identifiable by means of spectral analysis of their light (taking the form of a shift of the individual lines of the spectrum).

> *When dilute gases glow they produce a so-called line spectrum, in other words, a spectrum that contains only certain characteristic wavelengths, which show up in a spectroscope as shining lines.*[75] *Spectral lines were first studied by the physicist Joseph Fraunhofer (1787-1826). The spectroscope, an instrument for the spectral analysis of light, was invented by Gustav Kirchhoff (1824-1887) and Robert Bunsen (1811-1899), but this was not until 1859, six years after Doppler's death. It was spectroscopic measurement that led in the 20th century to a revolution in astrophysics and cosmology. It not only helped determine the chemical composition of the stars and nebulae,*[76] *but also enabled the measurement of the speed of star motion, precisely on the basis of the effect discovered by Doppler. The Doppler Effect has found a whole number of applications in almost all branches of science and technology—from measurement of blood flow to police radar.*

Due to the onset of cancer of the larynx, however, Doppler was obliged to restrict his teaching activities. The political climate was also worsening and in 1848 there was unrest in Prague. Doppler left Prague and worked for a short time at the Academy of Mining and Forestry at Banská Štiavnica in Slovakia, before returning to Vienna. He founded and directed the Institute of Physics and became a member of the Viennese Imperial Academy of Sciences and an honorary doctor of the Charles-Ferdinand University in Prague. He died in Venice at the untimely age of 49 and he is buried on the island of San Michele.

---

[75] Gases glow if their molecules are in an excited state. Excitation generally occurs due to raised temperature or the passage of an electrical current (discharge). At higher concentrations (higher pressure) the spectral lines widen until they join up in a continuous spectrum.

[76] In 1825 the outstanding French philosopher Auguste Comte (the founder of positivism and co-founder of sociology), still asserted that the chemical composition of the stars would never be ascertained. Fortunately, physicists did not take the philosopher seriously.

Secondary schools were named after Doppler in his native Salzburg and in Prague. His name was also given to asteroid 1984 QO, discovered by the Czech Antonín Mrkos. However, the greatest monument to the scientist is the Doppler Effect itself.

Arthur Stanley Eddington
(1882-1944)

The British astrophysicist Sir Arthur Eddington was the first to measure the bending of light rays caused by gravitation, and thereby helped achieve widespread acceptance of the general theory of relativity, which subsequently played the leading role in cosmology and in almost all of modern physics.

He was born in Kendal in the Lake District of northern England. He studied in Manchester and at Trinity College, Cambridge. In 1906 he was appointed to a post at the Royal Greenwich Observatory. He undertook a legendary expedition in May 1919 to the cocoa-producing island of Principe in order to observe the bending of stars' rays as they passed near the Sun's disc when it was in eclipse. These measurements were intended to test the predictions of Einstein's general theory of relativity. The widely-publicised results were excellent although the observation of the eclipse was fraught with many problems. Heat destroyed most of the photographic plates, leaving only eight usable ones. There was a rainstorm that lasted until just two hours before the eclipse and the sky was still overcast five minutes before. The observers did not even manage to abide by many of the methodological principles necessary for correct assessment of the measurements. Nevertheless there was a surprising correlation between the theory and the observation. (In 1983 British astronomers re-measured all the original photos using modern digital measuring equipment and the results showed an even better value of deviation, with a relative error of 7%.)

Eddington also made a significant contribution to astrophysics. He realised that the conditions obtaining in the stars were markedly different from those we are accustomed to in laboratories. Above all the pressure of radiation, which tends to be negligible on Earth, plays a much more important role, whereas in the stars the pressure of gaseous material is preponderant. More energy is carried through radiation than by convection. Eddington calculated that the temperature inside the stars must be as high as millions of degrees. He also deduced the relationship between the mass of the stars and their luminosity and explained the periodical changes in the luminosity of variable stars known as Cepheids.

> *Eddington lectured on relativist physics, and Einstein himself considered his book as the best textbook on relativity. The story goes that a journalist declared in Eddington's presence that the theory of relativity was so complicated that only three people in the world understood it. Eddington reflected for a moment and then asked him who the third was . . .*

From the 1920s until his death, Eddington worked on what he called the "fundamental theory" that was intended to be a unification of quantum physics, the physics of relativity, and the theory of gravitation. He speculatively combined various dimensionless ratios of fundamental physical constants in the belief that he would thereby discover the basis of his fundamental theory. He failed, however, and his approach is now regarded as pseudo-scientific.

> *Eddington's authority was also undermined by the reasons he gave for the value of the so-called fine-structure constant (which characterises the strength of electromagnetic interactions). According to experiments, its value was 1/136, which Eddington accounted for by his theory. When more precise measurement corrected the value to 1/137, Eddington quickly advanced a new theory to show why it should be precisely 1/137. This earned him the nickname of "Arthur Adding-One".*

Another unfortunate matter was Eddington's rejection of a theory by the young Indian astrophysicist Subrahmanyan Chandrasekhar (1910-1995), concerning the maximum mass of white dwarf stars.

(The mass of a star can increase by matter overflowing from the other component of a binary star system. If critical mass is exceeded the star collapses and a huge explosion occurs—supernova Ia). Chandrasekhar received a Nobel Prize in 1983 for his theory.

Eddington died in Cambridge. Asteroid No. 2761 and a crater on the Moon were named after him in his honour.

Albert Einstein
(1879-1955)

Einstein was born into a Jewish family in the German city of Ulm, although he soon moved with his family to Munich. He was no child prodigy. After overcoming his problems with grades, a reassuring fact for parents of underachievers, he passed his secondary-school leaving exam and was enrolled at the Zurich Polytechnic. He graduated with a diploma to teach mathematics and physics. After two years seeking in vain for work, he got a job at the Bern patent office, where he also dealt with theoretical physics. He published his special theory of relativity and other important papers. He then managed to obtain the post of professor of physics at the Zurich Polytechnic. In 1911 he arrived with his wife in Prague where he took up the post of professor of theoretical physics at the Charles-Ferdinand University. A year later he returned to Zurich and then moved to Berlin, where he took up an appointment at the Kaiser Wilhelm Institute (now the Max Planck Institute). Faced with the spread of anti-Semitic sentiment in Nazi Germany, Einstein left Berlin in 1933 for Princeton, New Jersey in the USA, where he taught at the famous university there until his death in April 1955.

Einstein's most important achievements include his theory of the photoelectric effect (the effect whereby electric current is released from the surface of metals by the action of light), which was published in 1905 and later helped create quantum mechanics. (It was chiefly for that work that he was awarded the Nobel Prize in 1921.) In that "annus mirabilis" of 1905, Einstein also published his special theory of relativity, which introduced a non-traditional concept of time and space. According to that theory, moving bodies grow shorter in the direction of motion (length contraction), and time in them flows more slowly (time dilation). In all inertial systems (i.e. coordinate systems governed by the law of inertia) the speed of light is the same and is unattainable for all bodies; at that speed a body would contract to zero, its mass would increase to infinity and it would have no time progression.

In 1907 Einstein came up with the principle of equivalence, according to which gravitational acceleration is indistinguishable from acceleration caused by mechanical force. The equivalence of gravitational and inertial mass also derives from this.[77] That same year he also formulated the principle of the equivalence of mass **m** and energy **E**, stated in the well-known equation $\mathbf{E = mc^2}$ (in which **c** is the speed of light in a vacuum). During his stay in Prague in 1911, he predicted the bending of light rays by a gravitational field and finally, at the end of 1915, he published his supreme work, the general theory of relativity.

The general theory formulates field equations that relate the density of mass to a space-time structure. According to the theory, light should travel in geodesic lines, whose curvature depends on a gravitational field. Einstein's prediction of the curvature of light was corroborated in 1919 by the expedition led by Arthur Eddington. Almost overnight, Einstein became a fixed star in the scientific firmament.

Not all of Einstein's efforts were crowned with success, however. For most of his life he tried in vain to come up with a theory that would integrate gravitational and electromagnetic interactions. Nor were his reservations about quantum mechanics accepted by the majority of physicists. (Einstein regarded it as an incomplete theory, as it was unable to provide precise predictions.)

---

[77] Although this principle is implicit in Newtonian mechanics, Einstein made it the foundation of his general theory of relativity.

Einstein is also renowned for his involvement in public life. He wrote a letter warning President Roosevelt about the threat posed by Nazi Germany's nuclear programme. Although a pacifist, he thereby hastened the development of the atom bomb. He was also involved in the Zionist movement, whose aim was to establish the state of Israel. He refused the offer to become president of that country, however.

Einstein's name and his distinctive physiognomy became a successful marketing icon. Einstein's name has been given not only to the chemical element einsteinium (atomic number 99), a satellite, Asteroid 2001, and also a range of commodities that have little in common with the physicist.

Epicurus of Samos
(341-270 BCE)

Epicurus was one of the most influential philosophers of Hellenistic Antiquity, i.e. the period after the death of Alexander the Great and Aristotle. He was born on the island of Samos. He pursued his career on the island of Lesbos before establishing his own school in Athens in 306 BCE.

As a boy he witnessed much human anxiety that was due to various superstitions. He therefore resolved to provide people with a philosophy that would not only satisfy their yearning for knowledge but would also rid them of unnecessary fears associated with the gods and suffering in the afterlife. He asserted that as long as the individual was healthy and not prey to irrational fears, pleasure would come of its own accord. This was because the gods did not look after people and death was no concern of humans. As long as one was alive death was not present and when it came one no longer existed. And after death there would be nothing, whether people liked it or not.

> *Epicurus consistently opposed myths and superstitions. He did not abolish the gods but did something even worse: he took away all their power. This soon had repercussions: slanders started to be spread about Epicurus and his school. (They gave rise to the later understanding of Epicureanism as total hedonism.)*
>
> *A derogatory squib was penned about Epicurus: fake "Courtesans' letters", in which the philosopher is portrayed as a pitiful and vain old lecher who relishes degenerate pleasures.*

Epicurus decided to build his philosophical outlook from the very foundations of physics, and specifically, ancient atomism. The Epicureans also made a significant contribution to the evolution of astronomical thinking. In Epicurus' view the universe was infinite and completely filled with stars. Human beings lived in an insignificant corner of the universe and at a time of no significance. They were no more than specks of dust and could only find their happiness within themselves.

Epicurus wrote three hundred works, thirty-seven of which were about nature. Present-day knowledge about Epicurus derives largely from the remarkable poem *On the Nature of Things* (*De Rerum Natura*) by Lucretius Carus.

Euclid of Alexandria
(c. 330-275 BCE)

Euclid became the embodiment of ancient geometry. He was born in Alexandria, Egypt, during the reign of Ptolemy I. He penned several works dealing with mathematics, astronomy and music, the most famous of which are his *Elements* (*Stoicheia*), consisting of thirteen books that set out all the known mathematical knowledge of that time.[78] Right up to the mid 19th century, Euclid's *Elements* was, after the *Bible*, the most published book in the world.

There is still controversy over Euclid's 5th postulate, which states: If a straight line crossing two straight lines makes the interior angles on the same side less than two right angles, the two straight lines, if extended indefinitely, meet on that side on which are the angles less than the two right angles. (There are various other equivalent wordings.)

Since the time of Euclid mathematicians have sought to prove this postulate. On many occasions it seemed that a proof had been found, but every time an error was detected—the logical fallacy of begging the question. It wasn't until the beginning of the 19th century that the world of mathematics realised that it was not a postulate but a distinctive axiom—the axiom of parallelism. It therefore cannot be deduced from other axioms, which means it also cannot be proven. This conclusion gave rise to speculation about whether geometry would be possible without that axiom. From there it was but a small step to the creation of non-Euclidian geometry. A small step, but a difficult one as it was necessary to abandon the preconception that only one geometry was possible.

Edmund Fournier Albe
(1868-1933)

Fournier d'Albe was a physicist, chemist, inventor and fantasist on a grand scale. He taught at the University of Birmingham and made a name through his work on the optoelectrical properties of selenium, becoming aware of the potential use of that semiconductor for photocells. He was the author of several inventions, including the optophone, an instrument to help blind people read printed text. In

---

[78] Much of the knowledge came from Euclid's predecessors, so one cannot agree with the assertion that "Euclid created geometry" (Steinhardt and Turok).

1923 he transmitted by television a portrait of King George, the first person to do so.

In a booklet *Two New Worlds* (1907), he voiced the hypothesis that the night sky was dark because the universe was dominated by dark, non-shining stars.[79] Were that true, there would have to be 10^{12} (one trillion) times more dark stars than shining ones. He also resurrected John Herschel's notion about the hierarchical structure of the universe. Fournier calculated that the gravitational potential at the surface of a sphere did not depend on its radius so long as the mass of the sphere is proportional to its diameter and not its volume. This situation could arise if cosmic matter was hierarchically distributed. (The density of such a system would come close to zero as the scale increased.) The model would solve also the paradox of darkness at night.

Fournier d'Albe also wrote various works of science fiction, which sounded scientifically plausible.

Alexander Alexandrovich Friedman
(1888-1925)

The Russian mathematician, physicist, meteorologist, aviator and adventurer Alexander Friedman (also spelled Friedmann or Fridman) came up with notable solutions to Einstein's field equations.[80] In so doing he removed cosmology from the sphere of the purely speculative sciences.

At St Petersburg University he devoted himself chiefly to meteorology. During World War I he became an aviator, organising

---

[79] The idea of dark stars had already been formulated by the French physicist, astronomer and pioneer of photography, François Arago. It appeared in his posthumous publication *Popular Astronomy* (1857).

[80] A system of ten non-linear partial differential equations. Their solutions are classes of functions.

the aerial navigation service and teaching at the school of aviation. After 1917 he served as temporary director of an aircraft factory in Moscow. In 1922 Friedman published his solution to Einstein's general relativity equations for a homogeneous distribution of matter.[81] It followed from this solution that the universe should either expand or shrink. Einstein rejected this conclusion, however, and did not even react to a letter from Friedman. He later recognised that the solution had been correct in mathematical terms and apologised to Friedman. Nevertheless Einstein persisted in his view that the universe was static and amended his equation by introducing the "cosmological constant" to correct it. Not long afterwards it was demonstrated that the universe really does expand. Friedman did not live to see his triumph, however; he suddenly fell ill with typhoid fever in August 1925 and died two weeks later at the age of 37.

Galileo Galilei
(1564-1642)

The turn of the seventeenth century was a period rich in giants of the human spirit and Galileo was foremost amongst them. He was baptized on the very day that Michelangelo passed away and two months before Shakespeare was born. He was a contemporary of such figures as Francis Bacon, René Descartes, Johannes Kepler, Tycho Brahe, Comenius and Otto von Guericke, as well as the ill-fated rebel Giordano Bruno. And Isaac Newton was born in the year of Galileo's death.

He made his name not only as an astronomer, but also as a mathematician, physicist, inventor and enlightened thinker, who helped Renaissance science break free of the restrictions of outdated Aristotelianism.

---

[81] A solution for a spherically symmetrical structure had already been found by Karl Schwarzschild (1873-1916).

Galileo's life story starts at Pisa, in Tuscany. He was forced to curtail his studies at the local university for financial reasons, but he was soon offered a post as a teacher of mathematics. He conducted experiments in search of physical laws. The book of nature, he said, was written in the language of mathematics, and this would be his model.

In 1604 Galileo endorsed Copernicus' heliocentrism. The Holy Inquisition had warned him against taking that step on the grounds that it directly contravened the *Bible* and the writings of Aristotle. The year 1609 was a watershed in the history of astronomy. That year Galileo made his first telescope (according to a description of a telescope constructed the previous year in Holland).[82] Galileo's telescope ("perspicillum") consisted of two lenses, a plano-convex objective and a plano-concave ocular. The first discoveries made with the help of the telescope were published in March 1610 in his treatise *The Sidereal Messenger*.

> *Perhaps Galileo's most important discovery described in the treatise were the four moons of Jupiter, a miniature Solar system. Jupiter's moons orbit the planet much faster than our Moon. They are a kind of "cosmic clock", and Galileo actually proposed their use for determining time in maritime navigation, although the idea turned out to be impractical. Moreover, this "cosmic clock" periodically loses or gains about twenty minutes. That phenomenon was later explained by the finite speed of light and the movement of the planets around the Sun. (See the section on Rømer.)*

Galileo's telescope showed our Galaxy—the Milky Way—to be made up of many small stars, as Democritus had supposed two thousand

---

[82] The patent application was lodged on 2nd October 1608 by Hans Lippershey (1570-1619), a German settled in Holland. It was perhaps Leonard Digges (father of Thomas Digges), who first constructed a functioning (reflecting) telescope sometime between 1540 and 1559.

years before. It was not a cloud, as was generally thought.[83] (However, Galileo supposed that all nebulae consisted of stars. Subsequently this opinion turned out to be incorrect.)

In spite of a warning from the Holy Inquisition Galileo published his famous work *Dialogue Concerning the Two Chief World Systems, Ptolemaic and Copernican*. The book is presented as a conversation between Simplicio, an advocate of geocentrism, Salviati, a defender of heliocentrism and Sagredo, an inquisitive layman. The heliocentric arguments prevail, of course.

> *In his Dialogue, Galileo sought to present among others, his theory of tidal forces. It derived from assumptions of heliocentrism and was intended to support that position. According to Galileo, high tides were caused by the addition of centrifugal forces as the Earth rotated and revolved around the Sun. According to that theory there should be a high tide every night. Galileo managed personally to convince Pope Urban VIII about his view. The pope was an old friend and sympathised with him, but he feared the consequences of a further proof of heliocentrism. He did not even notice that Galileo's theory was inconsistent with experience, but hastily advanced a theological argument: the theory was proof of nothing, because almighty God can cause any tidal phenomena that He pleases. Galileo then promised to include the pope's comment in his book. However, he placed the pope's argument in the mouth of the naïve Simplicio. He didn't get away with this, and the pope personally forbade the book's distribution.*[84]

---

[83] The stellar structure of other galaxies was discovered by Edwin Hubble in 1923.
[84] Kepler had already advanced a better theory, according to which tidal phenomena were caused by the Moon. He had written to Galileo about it, but the self-assured Galileo gave his younger colleague the brush-off.

Galileo discovered many interesting astronomical phenomena including the phases of the planet Venus, the strange elliptical shape of the planet Saturn[85], and dark spots on the Sun. Thanks to these spots he was able to discern that it rotated.

Galileo was summoned to Rome to face trial by the Inquisition. He was obliged to spend two whole months in prison awaiting trial (albeit the prison was a luxurious place: he had several rooms at his disposal, as well as good food and a servant). He was threatened and called on to "abjure, curse and detest" his views. As we know, in that difficult situation Galileo did not opt for the fate of Socrates, Jan Hus or Giordano Bruno. He had no yearning for martyrdom, so he preferred to comply with the judgement his judges commanded. Instead of the stake, all that awaited him was a life sentence of house arrest. He died blind in January 1642. According to a legend that is widespread but not historically substantiated, he pronounced before his death the memorable words "Eppur si muove" ("It moves nonetheless").

Galileo was buried in the Basilica of Santa Croce in Florence. His manuscripts ended up in the hands of a pork butcher who used the paper for wrapping his sausages. A customer recognised Galileo's handwriting and bought the paper for a song. The preserved writings are now deposited in the National Library in Florence.

It was not until October 1992 that pope John Paul II issued a ruling overturning the Inquisition's verdict on Galileo. He said that the Commission of the Inquisition had acted in good faith and in accordance with its level of knowledge. So it was not a controversy between science and the church, but a tragic mutual misunderstanding. An official apology came in 2000.
Galileo is commemorated by a lunar crater, a crater on Mars, asteroid 697 and the forthcoming(?) European satellite navigation system.

---

[85] Saturn's rings were recognized by Christian Huygens thirteen years after the death of Galileo.

George Gamow
(1904-1968)

It is due to Gamow that science has the Big Bang theory, and due to Stalin that America had Gamow.

Gamow was a Russian from Odessa. As a boy he lived through several world-shaking events: the return of Halley's Comet, the October Revolution, and the Civil War in Russia. He commenced his studies at Odessa University, but after a year he left for Petrograd (from 1924 Leningrad), where he would become a student of the famous Alexander Friedman. However, Friedman's death brought that to an end. As an outstanding student he was permitted to undertake post-graduate studies at Göttingen, where he obtained a doctorate. He spent three years at the Institute of Theoretical Physics at Copenhagen University and worked with Ernest Rutherford at the prominent Cavendish Laboratory at Cambridge, becoming a leading nuclear physicist. Communist propaganda extolled him as a prototype of the class-conscious Soviet citizen. On his return to the Soviet Union he worked at a number of establishments. However the Stalinist regime was becoming more repressive and, in spite of his prominent position, Gamow decided to leave the Soviet Union. This was a difficult and often very dangerous enterprise. Together with his wife he tried to cross the 250 kilometres of the Black Sea to reach Turkey, and on another occasion from Murmansk to Norway. His third attempt was less risky. In 1933 he and his wife travelled to Brussels for the prestigious Solvay Conference. When it was over they set off for the United States, where they settled and Russian Georgiy became American George. He took up at a post at George Washington University. This did not mean an end to paradoxes, however. While most American nuclear physicists were working on the atom bomb, Gamow, as an immigrant from Stalin's Soviet Union, was regarded as a potential spy. He failed to be positively screened, and was therefore free to concentrate on cosmology. He wrote and illustrated specialist and popular science books about physics. He died unexpectedly on August 19th, 1968 and is buried in the state of Colorado.

Gamow's major achievements include the "liquid drop" model of the atomic nucleus and the theory of the alpha decay. His key cosmological paper *The Origin of Chemical Elements* was published on 1ˢᵗ April 1948, although it was not an April fool's jape. Together with his colleague Ralph Alpher, he explained there the relative amount of hydrogen and helium in the universe as a result of the Big Bang.[86] A few months later he published together with Ralph Alpher and Robert Herman a paper that concluded that residual radiation (now called Cosmic Background Radiation or relic radiation) should have remained in the universe after the Big Bang. Gamow, then urged physicists to look for that radiation.

> *Residual radiation is thermal in nature. If a body is at a temperature of over 500 °C it emits visible light. At lower temperatures, radiation has a longer wavelength and is invisible (infrared radiation).*
>
> *The intensity of radiation is directly proportional to the fourth power of absolute temperature, i.e. temperature in kelvins (the Stefan-Boltzmann law). The wavelength of maximum emission of any body is inversely proportional to its absolute temperature (Wien's law). It follows from that when a radiating body is at a low temperature the intensity of radiation is very small and the wavelength of the radiation lies within the invisible zone of long waves. This also applies to residual radiation. At the time when the radiation was emitted it took the form of light. Now, 13 billion years later, it is microwave radiation, because as the universe has expanded its wavelength has increased.*

---

[86] The theory is known as the Alpher-Bethe-Gamow theory. By present-day standards the calculations were very rough, but the idea that the universe was originally very hot and the elements emerged as it cooled down has proved to be correct. Gamow added Bethe's name so that it would resemble the first letters of the Greek alphabet. Nevertheless Hans Bethe (1906-2005) wasn't just anybody: he was awarded the Nobel Prize in 1967 for his work on stellar nucleosynthesis.

For a long time astronomers did not take Gamow's theory seriously. It was Fred Hoyle, the renowned co-author of the universally accepted theory of a steady state universe, who derided Gamow's theory as the "Big Bang". That name superseded its original description of "theory of a dynamically evolving universe", which was more accurate but definitely weaker from a marketing point of view. Hoyle failed to see that it was Gamow's theory that would become the basis of modern cosmology, while his own steady state theory would soon end up in the imaginary museum of physics. Astronomers did not even make any great effort to search for the predicted radiation. In their defence it should be pointed out that methods for detecting microwaves were still in their infancy. And so residual radiation was not discovered until the 1960s, and only by chance. Its discoverers were awarded the Nobel Prize, an accolade that Gamow did not live to receive . . .

Thomas Gold
(1920-2004)

Thomas Gold had an extremely wide spectrum of interests. He worked in the fields of astrophysics, biophysics, geophysics and space-flight planning. He was one of the creators and protagonists of the steady state model of the universe.

In 1930 he moved from his native Vienna to Berlin. When Hitler came to power, Thomas' father—a Jew—left for Switzerland and then for England. Thomas attended school in Switzerland before joining his father in Britain. By a quirk of fate he found himself at the outbreak of war in the situation of an enemy alien. He had retained his Austrian citizenship and was therefore sent to an internment camp. There he met Hermann Bondi (likewise a Jew and an "enemy alien") and they became friends, later collaborating on the development of radar. They continued to work together in the spheres of astrophysics and cosmology. They developed the theory of the steady state universe, according to which the dilution caused by cosmic expansion was constantly compensated for by the creation of new matter.

Shortly after the discovery of pulsars (stars emitting regular pulses on radio frequencies) at the end of the 1960s, Gold and Hoyle developed a theory that these were rapidly rotating neutron stars with a powerful magnetic field. Their theory seemed so absurd that they didn't even allow Gold to present it at the astronomical congress. After the discovery of a pulsar in the Crab nebula, however, the situation changed and present-day astronomers accept Gold's explanation as an undeniable fact. Towards the end of his life, Gold postulated a theory that petroleum and coal were not of organic origin. History did not repeat itself in this case, and his views have not prevailed.

Thomas Gold died in the summer of 2004 in Ithaca, New York.

Otto von Guericke
(1602-1686)

A senior contemporary of Newton, the German Otto von Guericke, was a man of many parts—soldier, physicist, philosopher, and, for many years, Mayor of Magdeburg. He was born a year after the death of Tycho Brahe. He studied at the universities of Leipzig, Helmstedt, Jena and Leyden. He travelled around England and France before returning to Magdeburg in November of 1625 and became an official in the city administration. He had not lost his thirst for knowledge, however. In 1650 he constructed a vacuum pump and publicly demonstrated the properties of a vacuum, namely, the unexpectedly powerful force of atmospheric pressure and the fact that air is essential for burning and for the transmission of sound.

> *He conducted his most famous experiment in Magdeburg. He joined two bronze hemispheres together, with a leather seal between them and then pumped all the air out of them. Atmospheric pressure held the two hemispheres together so firmly that not even sixteen horses could pull them apart. But as soon as von Guericke opened the valve it gave a hiss and the ball collapsed of its own accord. The idea of a vacuum was revolutionary at that time. The ancient "logical proof of the*

> *non-existence of a vacuum" still survived—to put it crudely: "a vacuum is nothing and the existence of nothing is non-existence", along with Aristotle's "horror vacui", translated as "nature abhors a vacuum".*

In 1663 he built a machine with rotating sulphur balls that created an electrical charge. He used to demonstrate powerful electrical sparks for the entertainment and edification of the crowd. He also constructed a thermometer based on gas expansion, and also a water barometer. He registered the changes in the readings of these instruments in connection with the weather.

He was raised to the nobility in recognition of his services. He died in Hamburg in 1686, the year that Newton's *Principia* were published.

Edmond Halley
(1656-1742)

Halley is best known for the comet named in his honour. From the point of view of cosmology, however, his most remarkable contribution was his discovery of stellar motion.

He was born in London into the family of a soap maker. He was more interested in stars than in soap, however. He published his first scientific paper (on sunspots) while he was still a student at Oxford University. At the age of twenty he travelled to the island of St Helena to observe the sky of the southern hemisphere. After his marriage he settled in London. When the sky was clear he would spend his nights at the telescope, while cloudy evenings were spent making calculations at his desk. He summed up the results of his efforts in a theory that justified Kepler's laws of planetary motion. Elated, he set off for Cambridge to surprise Newton with his discovery. But the surprise was to be his—Newton had already solved the problem, but had not published his results. Halley persuaded him not to keep his discoveries to himself and he became Newton's friend and patron.

In 1698 he organised an Atlantic expedition to map the earth's magnetic field. Halley then tried to explain the variation of the magnetic field from the North-South line. He imagined the earth to be hollow and consisting of three concentric shells, inside which lay a core. The shells rotated at different speeds, and each shell had two magnetic poles. Halley even conceded that each of the shells might be populated. He held that their inner atmospheres were luminous and the escape of light through a polar opening in the outer spherical shell caused the Aurora Borealis.[87] Although Halley's explanation is more fantasy than science it does have a certain rational aspect—arranging the complex structure of the field into a sum of components that correspond to a simpler bipolar nature.

> *The hollow earth theory may seem absurd, but it was to enjoy considerable longevity. In 1818 it was revisited by an American, John Cleves Symmes, Jr. (1779-1829), a former army captain, who maintained that there were openings at both terrestrial poles giving access to four inner spheres. He proposed a military expedition to annex the territory inside our planet—in the name of the United States of America, of course. In 1870, the American Cyrus Reed Teed (1839-1908) made an even bolder modification of the theory. He claimed that human beings live on the inner surface of a hollow in an endless rock. Rather than proving his hypothesis scientifically he concentrated on preaching it to others and established a sect. That sect died out in time but the story of the hollow earth theory continued. It fascinated a German pilot by the name of Peter Bender, who was taken prisoner in France during World War I. While there, he developed the theory of a hollow Earth further and declared the curved surface and other proofs of Earth's spherical nature to be optical illusions. Bender's theory was taken up by the leadership of the Wehrmacht, which organised experiments in 1942 on the Baltic island of Rügen to see whether it was possible to observe British ships that appeared*

---

[87] It should be borne in mind that no human had yet set foot on either of the terrestrial poles.

> beyond the horizon by an optical illusion, whereas in fact they were above the horizon. The experiment was, predictably, a failure, and Bender himself ended up in a concentration camp.

In November 1703, Halley was elected to the chair of geometry at Oxford University. Two years later, he expressed his belief that the comets sighted in 1456, 1531, 1607 and 1682 related to the same cosmic body, and he predicted that it would return in 1758. The comet did indeed return that year and shone in the sky on Christmas Day, although Halley did not live to enjoy his moment of glory. Apart from comets, Halley also discovered in the constellation of Hercules the globular star cluster M13, a celestial gem that can be observed even with quite small telescopes.[88] And in 1718 he established that stars as well as planets moved through the sky. Their apparent velocity is substantially smaller, however.

Globular Cluster M13

---

[88] Globular cluster M13 in the Hercules constellation is 25 000 light years away. Its diameter is about 150 light years, i.e. twenty times more than the distance from Sirius to our planet. It consists of a hundred thousands stars; their concentration is about 500 times greater than in our vicinity. About 160 similar clusters surround our Galaxy.

Halley also turned his mind to practical matters. He improved the diving bell so that the air inside could be replenished. And thanks to his statistical methods the British government was able to sell life annuities.

Halley is buried in the graveyard of the old church of St Margaret at Lee in South London. Apart from the famous comet, Halley's name is commemorated by craters on the Moon and Mars, and by asteroid 2688.

Edward Harrison
(1919-2007)

Harrison was a British-born cosmologist who spent much of his career in America.

Born in London, he served in the British army during World War II. Following World War II he was involved in atomic energy research, and then he moved to the United States as a research associate with NASA. He taught at the University of Massachusetts and several other universities and scientific institutions.

In 1964 he formulated an explanation of nocturnal darkness, which stated that it was due to the lack of energy (energy density) in the universe:

THERE IS NOT ENOUGH ENERGY IN THE UNIVERSE TO CREATE A BRIGHT STARRY SKY.

He pointed out that even if all matter in the universe was transformed into radiation (according to Einstein's equation **$E=mc^2$**), the corresponding radiation temperature would be only 20 K, i.e.—250° C. The universe is simply very old and therefore bloated. Its energy is too diluted in space, and so the ancient glow of the plasma has turned into faint invisible radiation. This assertion is correct, of course. However, it emerged from one factor only. There could be a clearer answer including the whole cosmological model, as has been explained in the first part of our book. In 1987 Harrison published a book, *Darkness at Night*, dedicated to this problem.

Harrison had a wide range of interest from nuclear energy, galactic magnetism and pulsars to the diffusion of dust in interstellar molecular clouds. He is also known for his work on the development of galaxies from primordial fluctuations.

Harrison died on 29th January 2007.

Heraclitus of Ephesus
(c. 535-475 BCE)

Ephesus in Asia Minor is one of the most remarkable places in the history of mankind. One of the wonders of the world—the Temple of Artemis—once stood there, and St Paul and St John the Evangelist both preached there. Tourists and pilgrims are shown a house where the Virgin Mary is purported to have died. And it is also where the eccentric philosopher Heraclitus lived and worked.

"What little I could understand of his works was admirable; and what I could not make out I have no doubt was equally excellent." That is apparently what Socrates said of Heraclitus.

Heraclitus' work was obscure and full of riddles and allegories, and is therefore open to many different interpretations. What emerges from the fragments that remain is an emphasis on order, dynamism and processes, symbolised by fire and flux (*You cannot step into the same river twice* . . .). The universe, according to Heraclitus, was

. . . AN EVER-LIVING FIRE, KINDLING ITSELF BY REGULAR MEASURES AND GOING OUT BY REGULAR MEASURES.

Heraclitus is therefore sometimes associated with the concept of ekpyrosis, a cyclical process, in which the universe starts and ends in a great fire. However, it is possible that ekpyrosis was a later Stoic invention.

Heraclitus shunned human company, which could be why he had no direct pupils. His influence was considerable, nonetheless; he was a major influence on the Stoics. But the mainstream of Western thinking took a different course, from a concept of process to a static understanding of reality, the search for constants, invariants and the "elimination of time". In the field of cosmology this accounted for the lengthy survival of static cosmological models. As has been noted, Einstein himself initially supported the preconception of a static universe.

John Herschel
(1792-1871)

Sir John Frederick William Herschel, the greatest astronomer of the nineteenth century, inherited his profession from his father. He was born in England, in Slough, near Windsor. He became a founding member of the Astronomical Society, which shortly became the Royal Astronomical Society, and in 1831 he was made a Knight of the Royal Guelphic Order.

In addition to astronomy, his work encompassed mathematics, chemistry, botany and optics. He discovered ultraviolet radiation and also came up with the idea of contact lenses. He was also a photographic pioneer: concepts such as "positive" and "negative", as well as the very term "photography" were his brainchildren.

In astronomy he completed the work of his father—the cataloguing of objects in the northern stellar skies—and he extended it to the southern skies. He built several reflecting telescopes. He undertook a lengthy expedition with his family to southern Africa (the area of Cape Town), where he mapped the skies of the southern hemisphere. He was visited there by the young Charles Darwin, who was then engaged on his legendary voyage in HMS Beagle. Herschel discovered 525 nebulae and star clusters, as well as 3300 double stars. He named seven of Saturn's moons and four of the moons of Uranus.

From our point of view, it is important that John Hershel showed that the absorption solution to the dark night paradox was false and proposed a hierarchical model of cosmic structure. He contributed to astrophysics by overturning the theory that ordinary combustion was the source of stellar energy.

John Herschel was laid to rest in London's Westminster Abbey, adjacent to the tombs of Isaac Newtown and the kings of England. A crater on the Moon and a girls' school in Cape Town were named after him.

William Herschel
(1738-1822)

Frederick William Herschel was one of the most successful telescope builders and is regarded as the founder of stellar astronomy. He also discovered infra-red radiation.

Herschel came from Hanover in Germany. He was born into a musical family and became a member of a military band at the age of fifteen. When war was threatening, the young William left the army and went to England, where he eventually settled in the charming spa town of Bath. There he became Director of Public Concerts. He also taught and composed symphonies, concertos and chamber works. Herschel's interest in musical theory led him into the study of mathematics. And mathematics led him into astronomy. He became an enthusiastic astronomer and constructor of astronomical telescopes. On 13th March 1781 he discovered a new planet, which he named "Georgium Sidus" in honour of King George III. It was the first planet discovered by means of a telescope, and the first not have been known since Antiquity. It was later named after Herschel, and later still, in 1840, it was given its present name of Uranus after the Greek god of heaven. After that spectacular discovery he abandoned music and devoted himself entirely to astronomy. In rapid succession he discovered the moons of Saturn and Uranus, compiled a catalogue of nebulae and double stars, described the rotation of Saturn and its rings, and recorded changes in the polar ice-caps on Mars. He believed that our Galaxy was one of many similar stellar systems.

Herschel stated that he could see a ring around Uranus, but it was not confirmed by later observations. And then, in 1977, photographs taken by a space telescope showed a ring. Was Herschel's discovery just an illusion or were his telescope and his experienced eye so consummate? Or was the ring more apparent at that period? Did Herschel really discover Uranus' ring?

For curiosity's sake it is worth noting that Herschel believed the Sun to be a mountainous inhabited sphere surrounded by a thick layer of clouds that protected its inhabitants from the extremely hot glow emitted by the outer layer of radiant clouds.

In 1789 he built the biggest telescope of his day. It was 12 metres long and its metal mirror was 126 cm in diameter. On the very first night of its use he discovered another of Saturn's moons. Herschel was greatly assisted in his astronomical observations by his sister Caroline, who discovered eight comets herself.

William Herschel died in August 1822 at his observatory in Slough (now part of London). In 1963 the observatory was bulldozed to make way for an office block. He was buried in the family grave in the village of Upton.

A small crater on the Moon, an impact basin on Mars and asteroid 2000 are named after him, and the largest European telescope at the observatory on the island of Palma also bears his name. In addition, the Herschel Space Observatory is the most efficient space telescope, operating in infrared range of radiation.

Frederick Hoyle
(1915-2001)

Sir Fred Hoyle made his name not only as an astrophysicist and cosmologist but also as a writer of science fiction.

He came from the north of England and worked for most of his life at the Institute of Astronomy at Cambridge University. After World War II he founded and led the Institute of Theoretical Astronomy. He studied the theory of the origin of stars and, together with the German physicist Martin Schwarzschild (1912-1997), he modelled the evolution of giant stars. He studied the mechanism of nucleosynthesis in stars, i.e. the "combustion" of helium into heavier elements (the "carbon cycle"). In 1948, together with Hermann Bondi and Thomas Gold, Hoyle developed a creation model of a steady-state universe. In a radio broadcast, he coined the pejorative term "big bang" to describe the cosmological model of his rivals. He was also interested in the origin of the Solar System and the nature of quasars. (Quasars are very distant star-like objects that radiate a great deal of energy; astronomers believe them to be supermassive black holes into which surrounding material falls and shines intensely in the process.) Together with Alfred Fowler (1911-1995) Hoyle revealed the mechanism whereby heavier chemical elements are created in supernovae—collapsars.

Many of Hoyle's views had more in common with science fiction than serious science, and they were not generally accepted by the scientific community. This also applies to his conjecture that life was brought to our planet by a comet. He also promoted the controversial theory of "intelligent design". He gave his imagination free rein in his many books of science fiction, of which the best-known is probably the novel *Black Cloud.*

He died in 2001. Asteroid 8077 is named after him.

Edwin Powell Hubble
(1889-1953)

No other discovery had as much significance for modern cosmology as Hubble's discovery of cosmic expansion.

Hubble was born in the American state of Missouri. He studied mathematics and astronomy at the University of Chicago. He then spent three years at Oxford University, where (under pressure from his father) he studied law and Spanish. On his return to the USA he taught and also practised law.

A decisive moment in Hubble's life was his departure for Mount Wilson Observatory in California in 1919, soon after the world's biggest telescope, with a mirror diameter of 2.5 metres, was completed. Hubble used the giant telescope to study nebulae, and in 1923 he discovered that a "nebula in Andromeda" was actually a galaxy, and soon afterwards he identified many more galaxies. He searched for Cepheids[89] in nearby galaxies and used those stars to prove that the galaxies were at a much greater distance from us than astronomers imagined. A further famous discovery was in 1929, when he recorded that the red shift of galactic spectra was proportional to their distance. He then used the Doppler Effect to interpret the results as evidence that the galaxies were moving away. (We will probably never know what part his assistant Milton Humason played

---

[89] Variable stars whose absolute luminosity can be determined and serve as standard candles for measuring distances.

in that achievement.) Nowadays one generally speaks of cosmic expansion, but Hubble was initially more cautious and referred only to "apparent" galactic velocities (because other explanations of red shift were possible).

The reliability of Hubble's measurements was later called into question—as with every great discovery. The result was said to be not entirely convincing because of the great variance of values. It was even conjectured that Hubble already knew the theory of cosmic expansion and had "adjusted" his results to fit the theory. We now know that those variations were caused by the motions of the galaxies themselves (their "peculiar motions") which were added to the velocity of general recession. Initial measurements gave the constant proportionality between the rate of expansion and distance (Hubble's constant) as **H** = 500 km.s^{-1}.Mpc^{-1}. (Mpc is the symbol for megaparsec or one million parsecs: approximately 3.10^{22} metres.)

> *Later, serious errors were detected in determining galactic distances, and nowadays the value of **H** is much lower: **H** = (72±3) km.s^{-1}.Mpc^{-1}. This means that two galaxies that are one megaparsec apart move away from each other at a speed of 72 km/s. This relative velocity of cosmic expansion is imperceptible by terrestrial standards. If galaxies were one kilometre apart they would have moved apart by only a thousandth of a millimetre in thirty years! So it is not surprising that this movement is not directly observable.*[90]

Hubble only interrupted his work at Mount Wilson during World War II, when he served in the army as head of ballistics at the Aberdeen Proving Ground in Maryland. Shortly before Hubble's death, the five-metre telescope at Mount Palomar (also in California) was completed. For a long time that instrument took over the

---

[90] Neither objects held together by gravity (stars, planets and the galaxies themselves) nor solid bodies (held together by electromagnetic forces) keep track of that expansion.

position of the world's largest telescope. Hubble was given the honour of being the first person to observe with it. He died at the end of September 1953 at San Marino, California. At his own request, the place of his burial was kept secret.

In spite of his important discoveries, Edwin Hubble was not awarded a Nobel Prize. It is said that he hired an agent to help him receive the prize. However, at that time the Nobel Committee did not regard astronomy as part of physics. Hubble's name is borne by a parameter (constant)[91] defining cosmic expansion, by Asteroid 2069, by a crater on the Moon, and most notably by the Hubble Space Telescope, which has been circling our planet for the last two decades.

Milton Lasell Humason
(1891-1972)

Humason was a rarity among 20th-century scientists, being an amateur with no formal education, who came into astronomy via the back door. Nevertheless astronomy owes much to him, and above all the fact that, together with Edwin Hubble, he discovered the law describing the expansion of the universe.

He was born in Minnesota and finished his schooling at the age of fourteen. He loved the mountains and earned his living as a muleteer, carrying materials to the summit of Mount Wilson, where the biggest observatory in the world was being built. That activity decided his future, and in 1917 he became janitor at the observatory. During the day he would take care of the building and at night he was a voluntary assistant at the observatory. George Hale, the observatory's founder and director, spotted Humason's talent and offered him the post of astronomer. That was a hard pill for the scientific community to swallow, as Humason lacked any formal education, let alone in astronomy. However, he became a meticulous

---

[91] **H** is generally called the "Hubble constant". "Parameter" is a more general term that implicitly allows for **H** to change.

observer, photographing and recording spectra of distant galaxies, perfecting observational techniques. He also discovered a comet and only just failed to discover Pluto, an image of which fell on a defect in the photographic plate. He measured the red shift of 620 galaxies. In the eyes of many of his colleagues, however, he remained an uneducated mule driver and suffered many wrongs at the hands of management and individuals. Because of that rivalry, when Humason discovered the first Cepheids in M31 ("the nebula in Andromeda"), the astronomer Harlow Shapley wiped the marks off the negative, his motive being that the discovery called into question Shapley's theory that M31 was part of our Galaxy.

A crater on the Moon and asteroid 2070 are named after Humason.

Lord Kelvin (William Thomson)
(1824-1907)

The best-known Irish physicist, William Thomson, was born in Belfast. He acquired a university education at Cambridge and in the years 1846-1899 he held the chair of professor at the University of Glasgow. In 1851 he was elected a fellow of the Royal Society of London. Thomson was actively involved in the laying of the transatlantic cable between Europe and America and used it to send the first cable to America from Europe. For his work on that project he was knighted by Queen Victoria and later ennobled, adopting the title of Baron Kelvin of Largs, which refers to the River Kelvin that flows beneath the windows of Glasgow University. The river's name not only became his title, but was also used to designate a newly introduced temperature scale.

Kelvin's scientific activities encompassed many fields: thermodynamics,[92] hydrodynamics and electromagnetism, as well as various aspects of mathematics and technology. In the field of astrophysics he pointed out that stars could not shine for ever because

---

[92] The very term "thermodynamics" was introduced by Kelvin in 1849.

the law of the conservation of energy also applied to them. He also calculated that heat must be generated at the Earth's core or it would cool down within a few thousand years.[93] Kelvin was the author of one statement of the second law of thermodynamics (the law of entropy), and elaborated the third principle of thermodynamics (which states that absolute zero temperature is impossible to attain). The concept of absolute temperature and the scale named after Kelvin were introduced in 1848. Kelvin is buried near to Isaac Newton and John Herschel in Westminster Abbey in London

Kelvin is best known nowadays for the scale of temperature named after him. Temperate expressed in kelvins is called "absolute temperature". A degree on the Kelvin scale is equal to a degree Celsius (centigrade), but the scale is shifted so that it starts at absolute zero, so that 0 °C equals 273.15 kelvin and 0 kelvin is equivalent to −273.15 °C.

Johannes Kepler
(1571-1630)

The founder of celestial mechanics, Johannes Kepler (Keppler, Kepplerus, Kheppler, Kapler, Kapner) was a German mathematician, optician, astronomer and astrologer. He was active at the court of Rudolf II in Prague for a period of twelve years. It was in Prague that he formulated his three laws of planetary motion, which provided the foundations for Newton's law of universal gravitation.

He was born seven years after Galileo in the town of Weil, not far from Stuttgart. (Now "Weil der Stadt", a small town of some 20 thousand inhabitants.) His life was marked on the one hand by his profound interest in astronomy and theology, and on the other, by

---

[93] These calculations played into the hands of those who believed that the Earth was only a few thousand years old, based on figures from the *Old Testament*. The current view is that heat at the Earth's core is generated by radioactive decay.

the tribulations of being a protestant at a time of recatholicisation and the Turkish threat. Kepler showed an interest in the stars and in God from childhood, when his early experience of observing a comet and a lunar eclipse set him on the path to astronomy. He studied theology, but his idiosyncratic views were unfavourably received by his fellows and his superiors, although his superiors did take note of his mathematical gifts and offered him the post of professor of mathematics and astronomy at the protestant school in Graz. He was also appointed district mathematician, whose duties included issuing astronomical and astrological almanacs. However, as victimisation of non-Catholics increased, Kepler decided to leave Graz.

I WANTED TO BE A THEOLOGIAN, AND FOR A WHILE I WAS ANGUISHED. BUT, NOW SEE HOW GOD IS ALSO GLORIFIED IN ASTRONOMY, THROUGH MY EFFORTS.

from a letter to Mästlin

In February 1600, at the age of twenty-eight, Kepler arrived in the more religiously tolerant city of Prague, and became assistant to the renowned Tycho Brahe. These were to be his most productive years. He analysed Brahe's observational data and carried out numerical calculations. He drew up planetary tables based on Tycho's measurements and named them the *Rudolphine Tables* in honour of the emperor. When Tycho Brahe died in 1601, Kepler was appointed his successor.[94] However, Kepler's stay in Prague was not a happy one in terms of his private life. His wife Barbara and his son died there. In January 1612 Rudolf II died, and Kepler soon moved to Linz, where he was appointed court mathematician. He remarried and fathered a further seven children. After a short period in Ulm he became an advisor to Albrecht von Wallenstein, and in 1627 he

---

[94] He was employed at one third of his predecessor's salary and the court was particularly dilatory about his payment.

moved to his employer's estate at Sagan.[95] He was chiefly employed as an astrologer, however. The Kepler family's financial situation continued to be precarious. He was owed large sums by the emperor and by Wallenstein. In 1630 the imperial diet was convened at Regensburg and Kepler saw this as an opportunity to meet the emperor and obtain at least some of what was due to him. He set out for Regensburg on horseback, but the weather was bad and the journey took him almost three weeks. He arrived too late, chilled to the bone and in a fever. He took to his bed and died five days later from pneumonia. He was laid to rest in the cemetery of St Peter's church beyond the city bounds, since, as a protestant, he could not be buried within the city. That rest was to prove brief, however, as the Thirty Years War was raging and the cemetery was turned into a battlefield, as a result of which the astronomer's grave disappeared. After many mishaps, Kepler's manuscripts were finally bought by the Russian empress Catherine II and are now preserved in St Petersburg.

Kepler sought order in the cosmos, and regarded that order as harmony, in line with Pythagorean thinking. He demonstrated that the planets did not travel round the Sun in circular orbits, as Copernicus had supposed, but in ellipses with the Sun at one of the two foci (Kepler's First Law of Planetary Motion).

> *The elliptical orbits of the planets are only slightly eccentric and are not very different from circles (the exception being the orbit of Pluto, but Kepler was unaware of its existence; however, Pluto was recently expelled from the family of planets, and its eccentric orbit was part of the reason). Kepler's laws can also be applied to the motions of asteroids, comets and artificial satellites. They do not take into account the mutual attraction of planets, however.*

From his analysis of the orbit of Mars, Kepler deduced that the radius vector of a planetary orbit (i.e. the line connecting the Sun and planet) describes equal areas in equal periods of time. It follows

---

[95] Albrecht von Wallenstein acquired Sagan in 1627. It is now in Poland.

that the velocity of the planet is highest close to the Sun (perihelion) and lowest at its furthest distance from the Sun (aphelion). This deduction is known as Kepler's Second Law of Planetary Motion.

> *Kepler realised that what kept the planets in their orbits and prevented them from flying off into space must be a force emanating from the Sun. But what kind of force was it? The law of universal gravitation was still unknown, but since the ancient times people had been aware of magnetism. So Kepler inclined to the view that the planets were held in their orbits by a magnetic force. This idea came from one of Kepler's contemporaries, the English physicist and physician William Gilbert (1544-1603). According to Gilbert, the magnetic forces that turned the compass needle did not come from the North Star (Polaris) as was generally believed, but from the our planet itself, which was an enormous magnet. Now every school student knows that the Earth itself is a magnet. The physical nature of terrestrial (and planetary) magnetism is still a subject of study. The favourite view that terrestrial magnetism is caused by a magnetised iron core is incorrect. The temperature of the core is several thousand degrees, a temperature at which iron is not magnetic. More precisely, when iron is hotter than the Curie temperature (770 °C for iron), it ceases to be ferromagnetic and also loses permanent magnetization.*

As was the custom in those days, astronomers made their living chiefly from astrology. In Kepler's words:

ASTROLOGY, SUPPORTS HER WISE BUT NEEDY MOTHER, ASTRONOMY, FROM THE PROFITS OF A PROFESSION NOT GENERALLY CONSIDERED CREDITABLE.

Nevertheless Kepler also found some arguments in support of astrology—pointing above all to its empirical nature.

In his *Dioptrice* (1611), Kepler proposed a fundamental improvement on the astronomical telescope by replacing the negative lens eyepiece of the Galilean telescope with a positive, converging lens. The main advantage of this arrangement was a larger field of view and the possibility of fitting crosshairs or a measuring device. It is not known, however, whether Kepler actually constructed such a telescope. It was probably his poor eyesight that prevented the great astronomer from observing "with his own eyes".

Kepler's grave has disappeared, but his name lives on. Craters on the Moon and on Mars and asteroid 1134 are named after him, as well as a university in Linz and some fourteen schools in various parts of the world, including a secondary school that stands near Prague Castle on the spot where Kepler lived.

Titus Lucretius Carus
(c. 97-55 BCE)

Lucretius was one of the foremost Roman representatives of Epicureanism. His life is shrouded in mystery, and his death was supposed to have had something to do with a love potion secretly administered to him. He left just one work, the didactic poem *On the Nature of Things (De Rerum Natura)*. Lucretius drew on Epicurus' teaching and dealt with everything to do with nature and man. We can find there views about the gods, explanations of physical events, descriptions of the movement of atoms and even proofs of the limitlessness of space. Lucretius did not limit himself to pure science, the poem also contains practical tips in the field of eroticism and sex.

Lucretius' poem indicates that Epicurean physics was much closer to modern concepts than that of Aristotle. It hints at certain laws that were not discovered until sixteen centuries later, by Galileo, such as the speed of a body's fall being independent of its mass. Lucretius also maintained that there were many worlds that are made and unmade all the time. Our world would also end one day. Nature was badly organised and full of evil, and man was the most unfortunate of beings in the universe. The human body and soul formed an inseparable whole, so the idea of reincarnation was nonsense, and that also applied to the existence of the soul apart from the body.

The poem concludes with attempts (sometimes naïve, from the present-day standpoint) to explain various natural phenomena: rain, rainbows, snow, harmful substances, miraculous springs, the Nile floods, magnetism, lightning, earthquakes, volcanic eruptions, and even a plague epidemic in Athens.

Lucretius' poem was already appreciated in Antiquity. It became a source of inspiration particularly for the Enlightenment philosophers of the 18th century, and was translated into all major languages.

Johann Heinrich von Mädler
(1794-1874)

One of the most important astronomers of the 19th century, the German scientist Johann Mädler, helped popularise astronomy in various ways.

He came from Berlin. Together with his friend Wilhelm Beer, a rich banker and an amateur astronomer, he established a private observatory in the grounds of the Berlin Zoo. They mapped the Moon and Mars and gave names to the features on the surface. Mädler also established the rotation time of Mars, with an error of only 13 seconds.

In 1836 Mädler was offered a position at the Berlin Observatory, and a year later he was appointed Professor of Astronomy. In 1840 he moved to Tartu in southern Estonia to run the observatory there. He worked particularly on the question of double stars and on measuring the proper motions of the stars. In his *Popular Astronomy* he also mentions a solution to the dark sky paradox. Initially he favoured the absorption solution, but in the last edition of 1869 he supported solution by the finite age of the universe. Craters on the Moon and on Mars were named after him.

Benoît Mandelbrot
(1924-2010)

Benoît Mandelbrot, a Franco-American mathematician of Jewish origin, has been called the "father of fractals". He was born in Warsaw, but when he was twelve, his family fled to France to escape the growing Nazi persecution of Jews. There he graduated from the Ecole Polytechnique after World War II. He then studied aeronautics in California. In 1955 he moved to Geneva, but in 1958 he returned to America, where he worked for IBM.

From 1951 he published papers in various areas of mathematics, information theory, economics and fluid dynamics. Based on his analysis of cotton price distribution he defined the concept of "fractal distribution" and conceived the theory of fractals. These geometrical structures have more in common with nature than the artificial smooth objects of traditional geometry. They occur everywhere: in the shape of mountains, coasts, lunar craters, various plants, pulmonary vessels, and also in groups of galaxies.

Mandelbrot's name was given to asteroid 27,500.

Isaac Newton
(1643[96]-1727)

Newton is the founder of modern (now "classical") physics. He discovered and mathematically formulated the basic laws of mechanics, and created an important mathematical instrument—infinitesimal calculus. He also built a simple type of reflecting telescope, which marked a turning point in astronomy.

---

[96] According to the Julian calendar that was then in use in Britain, Newton was born on 25th December 1642. According to the Gregorian Calendar, which we use nowadays, he was born in 1643.

He was born in Woolsthorpe (about 170 km north of London). He studied at Trinity College, Cambridge, obtaining his degree in the spring of 1665. An outbreak of the plague in the summer of that year caused the university to close, and the young graduate returned home. It was during that plague-imposed vacation that Newton made his most important discoveries. The story of the apple said to have fallen on the scholar's head dates from that period.[97] In 1667 he returned to Cambridge, where he was awarded a Master of Arts degree, and two years later he was appointed Professor of Mathematics.

Unwarranted delays in publishing his work inconvenienced Newton on several occasions. In 1679, for example, the physicist Robert Hooke wrote him a letter setting out the relationship between planetary orbits and solar attraction. Newton did not react to the letter, as a result of which Hooke assumed that it was he, Hooke, who had first discovered the law of gravity. A fierce dispute also arose concerning the discovery of differential and integral calculus, which had also been claimed by the German mathematician and philosopher Gottfried Leibniz (1646-1716). Eventually the British Royal Society ruled in Newton's favour. That historic meeting was chaired by Newton himself. Newton's key work *Mathematical Principles of Natural Philosophy* (*Philosophiae Naturalis Principia Mathematica*) was published in 1687. In 1688 Newton built a simple reflecting telescope. He used a ground metal mirror with a diameter of only 35 mm as its objective. Telescopes of the Newton type continue to be used in astronomy. They do not suffer from the chromatic aberration that causes coloured contours, and thanks to their simple construction they offer maximum efficiency at lowest cost.[98]

Newton was not only a physicist of genius, he was also a person of deep belief. However, he had an individual view of religion that led to disputes with official circles. He attempted to provide dates for

---

[97] That blow to his head might have been subsequent poetic licence on the part of Voltaire.
[98] Another advantage is that mirrors can be manufactured in larger dimensions than lenses.

ancient historical events, including the actual creation of the world, by analysing the *Old Testament*. This aspect of Newton is in sharp contrast with the traditional schoolbook version of his personality.

> *Newton's esoteric writings first came to light after World War II. They had been bought at auction by the world-famous economist John Maynard Keynes in 1936, but Keynes decided to postpone their publication until after the war. Keynes died before achieving this aim, and the papers were published by his brother.*

Isaac Newton died at the age of 84 and was buried in Westminster Abbey. The unit of force is now named after him. (One newton–1 N–is equal to the force that would give a mass of one kilogram an acceleration of one meter per second squared.) The temperature scale devised by Newton (°N) is no longer used.

Heinrich Wilhelm Matthäus Olbers
(1758-1840)

The German astronomer Heinrich Olbers was without doubt a remarkable personality. He was born at Arbergen, near Bremen. A doctor of medicine by profession, he spent his evenings observing the starry universe. After the death of his daughter and wife he gave up medicine and devoted the last twenty years of his life entirely to the study of the universe.

He devised the first practical method for calculating the orbits of comets, eighteen of which he calculated himself. In 1815 he discovered a comet with an orbital period of 72 days, which now bears his name. He also discovered two asteroids: Pallas and Vesta. He was of the view that infinite space is probably filled with stars accompanied by planets and comets. And he was convinced that the Moon had once been inhabited by intelligent beings.

Nowadays, however, he is known chiefly for the dark night sky paradox that he popularised in his articles. (However, he did not discover it, nor did he provide a correct explanation for it.) Not only the "Olbers Paradox" and a comet were named after him, but also a crater on the Moon.

Arno Allan Penzias
(b. 1933)

The joint discoverer of cosmic microwave background radiation, the American physicist Arno Penzias, was born in Munich. He arrived in Britain just before the outbreak of World War II as part of the Kindertransport operation that rescued Jewish children from Nazi Germany. He moved with his parents to New York in 1940. He graduated from the City College of New York and Columbia University, and in 1961 became an employee of Bell Laboratories, where he worked on the development of ultra-sensitive cryogenic microwave receivers. He was also involved with radio astronomy and became the director of the Radio Research Laboratory.

He discovered cosmic microwave background radiation jointly with Robert Wilson, and both scientists were awarded the 1978 Nobel Prize for their achievement.

Philolaus of Croton
(c. 470-385 BCE)

The Pythogorean Philolaus lived in the city of Croton in Calabria (Southern Italy) and was a contemporary of Socrates. His writings influenced Plato and Aristotle. His main interest was the use of mathematics to calculate the orbits of celestial bodies. He was possibly the first to use the term "cosmos" in its modern meaning of "universe" (it had previously meant "order" or "ornament"). He devised a remarkable non-geocentric system, in which the planets (including the Sun and Earth) revolved around a kind of central fire—a "cosmic hearth". The Sun was deemed to be alive and to transmit the divine fire by reflecting the light from the hidden central fire. The idea that celestial objects—particularly planets—were alive and were possessed of souls, was a widespread notion in the Middle Ages and it survived into the Renaissance.[99]

Edgar Allan Poe
(1809-1849)

The American writer and poet Edgar Allan Poe was the author of stories of mystery and fantasy. In addition, his poem *Eureka* offers a remarkable explanation of the dark night sky paradox

Born in Boston, he was orphaned at the age of three and was taken into the family of the merchant John Allan, living with them in Britain for five years. After returning to America he studied literature at university, but fell prey increasingly to alcohol and gambling. He attempted an army career but was dismissed from the military academy on grounds of lack of discipline. He experienced brighter moments in his life, however, when his literary works gained an enthusiastic reception. In 1834 he was awarded first prize in a literary competition, and a year later he was appointed editor of a leading literary journal. In 1845 he won fame with his poem *The Raven* and in 1848 he penned *Eureka: a Prose Poem*. In this poem, he mentions the astronomical discoveries of his day and seeks to find solutions to unsolved problems in his own way—mostly with a dilettante approach. Nonetheless, he indicated the right course to follow to solve the dark night sky riddle.

Poe never overcame his harmful addictions. In October 1849 he was found in a serious condition in the street in Baltimore, and he died soon after. Was he sick, drunk or high on drugs? He was just forty years old.

---

[99] Even in our own day, James Lovelock has advanced the theory of Gaia—the living planet. And Lee Smolin considers the entire universe to be alive. This raises an interesting issue about what is meant by "being alive".

Why is it dark at night?

Claudius Ptolemy
(c. 85-165 CE)

The Greek astronomer Claudius Ptolemy put the final touches to the geocentric system of Aristotle and Hipparchus.

He lived at Alexandria in Egypt, but we know almost nothing about his life. However, it is known that he did not belong to the ruling Ptolemaic dynasty, despite what many have assumed on account of his name. Apart from observing the stars, he also engaged in experiments, particularly in the field of optics. According to Ptolemy the Earth was at the centre of the universe. The planets, including the Sun and the Moon, revolved around the Earth. The universe was encased by the sphere of the fixed stars. The planets performed complex motions, moving in small circles (epicycles), which in turn move along a larger circle (deferent), with the motionless Earth at the centre.[100]

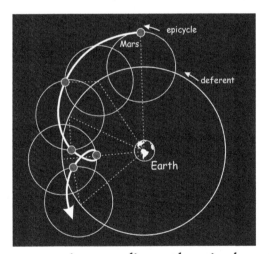

Planetary motion according to the epicycle model

The Ptolemaic system was also adopted by medieval scholasticism and by the Christian catechism. It was particularly suitable because of its

---

[100] From a heliocentric perspective, the epicycle is a projection of the Earth's orbit—and thus has the same origin as stellar parallax.

geocentrism, which was in line with the letter of the *Old Testament*. It was not fundamentally called into question until Copernicus and other heliocentrists, and it was convincingly refuted by Kepler.

Ptolemy's main astronomical work is known under its arabised title of *Almagest*. It consists of a catalogue of 1025 stars and mentions 48 constellations. Ptolemy added to the *Almagest* four books of astrological interpretations (*Tetrabiblos*), and he also wrote a popular version under the title *Planetary Hypotheses*. He also wrote a treatise on music and eight books about cartography, which give the geographical coordinates of eight thousand locations.

Ptolemy's name has been given to a large crater at the centre of the visible hemisphere of the Moon.

Georg Friedrich Bernhard Riemann
(1826-1866)

The German mathematician Bernhard Riemann was one of the fathers of non-Euclidean geometry, which was later used in relativistic physics and cosmology.

Insidious illnesses clouded Riemann's early years. Both his parents died of TB and the disease gradually claimed the lives of his siblings also. The young Georg wanted to become a Lutheran pastor like his father, but later his interest in mathematics prevailed. He studied at the University of Göttingen under the renowned mathematician Carl Gauss (1777-1855). It was Gauss who, in 1799, had discovered that geometry was possible without Euclid's fifth postulate—the parallel postulate. (Gauss found this so disconcerting that he did not even publish his discovery.) The idea of a non-Euclidian geometry was put forward a quarter of a century later by the Russian Nikolai Lobachevsky (1792-1856) and the Hungarian János Bolyai (1802-1860). Riemann generalised their findings and demonstrated that curved space can have any number of dimensions. Sixty years later, Einstein would formulate his general theory of relativity on the basis of Riemann's work: that theory describes gravity precisely by means of spatial curvature (or more precisely, space-time curvature).

In 1859 Riemann was appointed to the chair of mathematics at Göttingen. Because of his worsening tuberculosis, he often made trips to sunny Italy, the last of them in 1866. On that occasion, he walked there as there were no trains running due to the Austro-Prussian war. His life's journey came to an end by the side of the beautiful Lake Maggiore. He was only forty years old.

Ole Christensen Rømer
(1644-1710)

The Danish mathematician and astronomer Ole Rømer (Olaf, Olans, Römer, Roamer) was the first to prove that light travels at a finite speed.

Born at Åarhus, he studied mathematics and astronomy in Copenhagen. During the 1670s he was sent, along with Jean Picard, to measure the position of Tycho Brahe's former observatory Uranienborg on the island of Hven. Together they observed about 140 eclipses of Jupiter's moon Io. Meanwhile, these eclipses were being observed at the observatory in Paris by Giovanni Cassini (1625-1712). The motion of Jupiter's moon thus served as a universal cosmic clock (as Galileo had once proposed). The measured data were then used to calculate the difference in longitude between the Uranienborg and Paris observatories. The two astronomers noticed an interesting phenomenon: the time intervals between the eclipses of Io became shorter when our planet was approaching Jupiter and longer when it was receding.

Cassini failed to draw the appropriate conclusions from those observations, however. This was eventually done by Rømer, whose data were then used by Christian Huygens to calculate the speed of light.

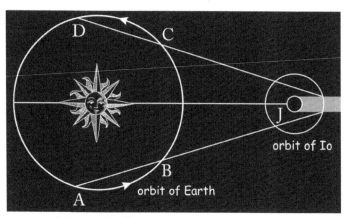

Measurement of the speed of light. Rømer compared time intervals between eclipses of Jupiter's moon Io when the Earth was approaching Jupiter (AB) with time when it was moving away from Jupiter (CD).

> *The idea that light travelled at a finite speed was not generally accepted until 1727, when the English astronomer James Bradley (1693-1762) measured what is known as aberration of light. Aberration is caused by a combination of the speed of light and the speed of the Earth in its orbit. This is because it takes a moment for light to pass through the inside of the telescope from the objective to the eyepiece. Our planet is not stationary, however, but rotates around its axis and revolves around the Sun. Therefore, as it travels through the telescope the ray of light is displaced slightly from its path. (Bradley measured the angle of aberration as 20.44 arcseconds; a more precise value is 20.4955 arcseconds.) We can get a visual idea of aberration when travelling in a car in the rain. Although the raindrops fall vertically, they travel obliquely in relation to the direction of the vehicle. Hence the front windscreen tends to be wetter than the one at the back.*

From 1672 Rømer worked at the Paris observatory and also as a tutor at the court of Louis XIV. He became a member of the French Royal Academy of Science. In 1681 the King of Denmark invited him to become his court astronomer, as well as professor of mathematics and astronomy, and director of the observatory. In 1700 Rømer persuaded the king to adopt the reformed Gregorian calendar. He introduced the first national system of weights and measures, as well as his own temperature scale, which is no longer used.

Rømer also became deputy chief of police in Copenhagen and took many measures to raise the city's standards. He introduced street lighting to enable control of beggars, the homeless and prostitutes. In 1705 he became mayor of Copenhagen and a year later, Chairman of the Danish State Council. He died shortly before his sixtieth birthday.

The Römer Crater is located in the north-east section of the Moon.

Harlow Shapley
(1885-1972)

The American astronomer Harlow Shapley measured distances to globular clusters, and determined the dimensions of our Galaxy and the location of our Solar System within the Galaxy. As late as the 1920s Shapley defended the theory that our Galaxy contained all the stars of the universe.

He was born in Nashville, Missouri. He wanted to become a journalist but the relevant university department had not yet been opened and so he opted for astronomy. He eventually ended up at the Mount Wilson observatory. His work there focused on Cepheids (periodic variable stars). He proved that their luminosity varied as a result of pulsation, and deduced the relationship that can be used to determine their absolute luminosity on the basis of measured periods of pulsation. Since then, astronomers have used Cepheids as standard candles for determining distances in the same way that mariners estimate distance by the intensity of the light of a lighthouse.

As a result of an error in measuring the distance of object M31 ("the nebula in Andromeda"), Shapley concluded that the "spiral nebulae" were also located in our Galaxy. Thus the material universe would consist solely of our Galaxy and the near surroundings, and beyond that there was a total void.

Shapley also had more earthly interests. His hobby was myrmecology—the study of ants. Shapley gave his name to a crater on the Moon, asteroid 1123, a planetary nebula and a supercluster of galaxies in the Centaurus constellation.

Robert Woodrow Wilson
(b. 1936)

Robert Wilson was born in Houston, Texas, now the centre of American space research. He studied physics at Rice University, Houston, and as a post-graduate at the California Institute of Technology at Pasadena. From 1963 he worked at the Bell Laboratories.

Thanks to a lucky accident, he and Arno Penzias discovered cosmic microwave radiation, which corresponded to the radiation of a black body at a temperature of about 3 K. They ascertained that the intensity of the radiation was independent of time; it was isotropic and was not polarised.

In 1978, Wilson and Penzias were awarded a Nobel Prize for their discovery.

Bibliography

Alpher, R. A., Bethe, H., Gamow, G., The Origin of Chemical Elements, *Phys. Rev. 73*, 7 p.803-804, 1948 (http://www.snolab.ca/public/JournalClub/michael1.pdf)

Augustinus Aurelius, *Confessiones (Confessions)*, (http://www9.georgetown.edu/faculty/jod/latinconf/latinconf.html)

Barrow, J. D., *The Origin of the Universe*, Orion, London 1994

Bondi, H., *Cosmology*, Cambridge University Press, Cambridge 1960

Diogenes Laertios (DL), *Lives and Opinions of Eminent Philosophers*, (http://ebooks.adelaide.edu.au/d/diogenes_laertius/lives_of_the_eminent_philosophers/contents.html)

Engels, F., *Dialectics of Nature*, (translated by C. Dutt), International Publishers, New York 1940 (http://www.marxists.org/archive/marx/works/1883/don/index.htm)

Galilei, G., *Dialogo sopra i due massimi sistemi del mondo (Dialogue Concerning the Two Chief World Systems)*, 1632 (http://www.liberliber.it/mediateca/libri/g/galilei/dialogo_sopra/pdf/dialog_p.pdf)

Halley, E., An Account of the cause of the Change of the Variation of the Magnetic Needle; with an Hypothesis of the Structure of the Internal Parts of the Earth, *Philosophical Transactions of Royal Society of London*, 195, p. 563-578, London 1692

Harrison, E., *Cosmology*, Cambridge University Press, Cambridge 2000

Harrison, E., *Darkness at Night*, Harvard University Press, Cambridge, Massachusetts, London, England 1987

Harrison, E., The Dark Night-Sky Riddle: A "Paradox" That Resisted Solution, *Science*, 226, 4677, p. 941-945, 1984

Hubble, E., A Relation between Distance and Radial Velocity among Extra-galactic Nebulae, *Proceedings of the National Academy of Sciences*, Vol. 15, Nu. 3, March 1929 (http://antwrp.gsfc.nasa.gov/diamond_jubilee/1996/hub_1929.html)

Kepler, J., *Kepler's Conversation with Galileo's Sidereal Messenger* (translated by Edward Rosen), Johnson Reprint, New York 1965

Kepler, J., *Somnium (The Dream)* (http://www.johanneskepler.info/index.php?option=com_content&view=article&id=43&Itemid=47)

Kirshner, R. P., *The Extravagant Universe: Exploding Stars, Dark Energy, and Accelerating Cosmos*, Princeton University Press 2002

Copernicus, N., *De revolutionibus orbium coelestium*, (http://www.bj.uj.edu.pl/bjmanus/revol/titlpg_e.html)

Koyré, Al., *Du monde Clos à l'Univers infini*, trad. Raïssa Tarr. Paris 2003 *(From the Closed World to the Infinite Universe)*, Publ. Johns Hopkins University Press 1968 (first published 1957)

Mädler, J. H., *Der Wunderbau des Weltalls, oder Populäre Astronomie*, Berlin 1861

Mandelbrot, B., *The Fractal Geometry of Nature*, W. H. Freeman and co., New York 1977 (updated 1983)

Newton D., *Olbers' Paradox, A Rewiew of Resolution to this Paradox*, http://astronomy.nmsu.edu/nicole/teaching/ASTR505/lectures/lecture29/olbers.pdf

Overvin, J. M., Wesson, P. S., *Light Dark Universe*, World Scientific Publishing Co. Ptc. Ltd., Singapore 2008

Penzias, A. A., Wilson, R. W., A measurement of excess antenna temperature at 4080-Mc/s, *Astrophys. J.*, 142, p. 419, 1965 *(http://adsabs.harvard.edu/doi/10.1086/148307)*

Poe, E. A., *Eureka, the prose poem* (http://xroads.virginia.edu/~hyper/poe/eureka.html)

Smolin, L., *The Life of the Cosmos,* Oxford University Press 1997

Steinhardt, P. J., Turok, N., *Endless Universe: Beyond the Big Bang*, Doubleday 2007

Tipler, F. J., Johann Mädler's resolution of Olbers' Paradox, *Quarterly Journal of the Royal Astronomical Society*, September 1988

Tipler, F. J., Olbers's Paradox, the Beginning of Creation, and Johann Mädler, *Journal of the History of Astronomy*, XIX, p. 45-48, 1988 (http://adsabs.harvard.edu/abs/1988JHA...19...45T)

Titus Lucretius Carus, *On the Nature of Things* (*De Rerum Natura*), R. E. Latham, transl., Penguin Books, London 1994 (http://www.gutenberg.org/ebooks/785)

Weinberg, St., *The First Three Minutes*, (1977 updated 1993) (http://kat.ph/15-steven-weinberg-the-first-three-minutes-a-modern-view-of-the-origin-of-the-universe-1977-pdf-t5666318.html)

About the Author

Peter Zamarovský (1952) comes from Prague, a city with a rich history of Czechs, Germans and Jews, a city where the famous astronomer Kepler made his most significant observations and where Danish astronomer Tycho Brahe found his last rest, a city in which Einstein started to formulate the ideas underlying the general theory of relativity. Inspired, perhaps, by the spirit of the place, Zamarovský studied physics at Charles University, and he now lectures in philosophy and physics at the Czech Technical University in Prague. He is chairman of the European Cultural Club, which, for over twenty years, has organised public panel discussions on topical issues related to various fields of science, philosophy and art. He is active in the historical section of the Czech Astronomical Society and in the Sisyphus Sceptics Club. Zamarovský has written many scientific and popular papers and three books. He also offers popular lectures on various themes in science and philosophy. He has been invited to participate in numerous radio and TV programs. His favourite saying is Hans Reichenbachs "The path of error is the path of truth."

Subject Index

A

aberration of stars (of light) 156
absolute (temperature) – see Kelvin scale
absorption interstellar 40, 45, 46, 47, 57, 64, 135, 147
accretion 68, 69, 70, 72
aging of photons 79-82, 92
anachronism 19, 20
apeiron 18
aphelion 145
Aristotle model (cosmos, universe) 29, 30, 90, 97, 104
astrology 18, 56, 145
astrophysics 20, 67, 69, 112, 113, 127, 135, 141
atomic nucleus 69, 79, 86, 126
atomism 21, 33, 108, 118
axiom of parallelism 119

B

big bang 74, 75, 82, 88, 90, 125-127, 137
black body 88, 89, 158

C

Celsius (centigrade) scale 142
central fire of cosmos 17, 151
Cepheid (variable stars) 56, 114, 138, 141, 157
collapsar 71, 137
collapse of star 71, 115
collisions of galaxies 54
collisions of stars (worlds) 19, 53, 54
combustion chemical 68, 135
Concordance model 86
cooling of radiation (of particles) 87
cosmic background radiation, CBR (= relic radiation) 88, 89, 100, 126, 151
cosmogony 17, 22, 63
cosmography 17
cosmological constant 76-78, 121
cosmological principle 23
curvature (of space) 76, 116, 155

D

deferent 153
determinism 109
Doppler Principle (shift, effect, interpretation) 77, 81, 92, 111-113, 138
double stars 49, 110, 135, 136, 147
dust (interstellar) 45, 47, 52, 55, 69, 71, 78, 85, 86, 118, 133

E

eclipse 15, 100, 103, 113, 143, 155, 156
egocentrism 90
ekpyrosis 26, 60, 134
electron 79, 86, 87

entropy 20, 48, 142
epicurean model (universe) 21, 23, 26, 28, 33, 36, 40, 41, 43, 44
epicycle 30, 153
ether (aether) 58
expansion of the universe 76, 78, 79, 81-83, 85, 89, 92, 127, 138-140
extinction (interstellar) 44

F

fluctuation 133
fractal 50, 51, 102, 148

G

galactic cannibalism 54
geocentrism 16, 17, 31, 90, 97, 101, 123, 154
geometry non-Euclidean (curved space) 79, 119, 154
globule 70
gravitation 2, 15, 41, 42, 52, 67, 75, 85, 86, 113, 114, 116, 142, 145

H

heliocentrism 34, 36, 93-95, 103, 106, 109, 122, 123
hollow Earth 130
horizon (cosmological) 65, 83
horror vacui 129
Hubble constant (parameter) 140
hydrogen 70-73, 86, 87, 126

I

infrared (radiation, range, spectrum) 70, 87, 126, 137

intelligent design (ID) 138

K

Kelvin scale, kelvin (degree Kelvin) 88, 126, 141, 142

L

logos 12, 14, 25

M

magnetism 133, 145, 147
materialism 2
myth 12, 13, 14, 31, 62, 118

N

nebulae 47, 55, 56, 61, 69, 73, 103, 112, 123, 128, 135, 136, 138, 157, 158
neutron star 71, 128

O

orbit 33, 34, 51, 67, 72, 96, 107, 144, 145, 149, 150, 153, 156

P

parallax of stars 41, 56, 94, 95, 96, 101, 102, 106, 153
parsec (Pc) 51, 72, 96, 139
perihelion 145
plasma 86-88, 90, 132
Prime Mover 97
proton 70, 76, 87
protostar 69
Ptolemy model 30, 153

pulsar 128

Q

quantum physics (mechanics) 90, 114-116

R

recombination 86, 87
red shift 77, 78, 79, 83, 85, 138, 139, 141
relativistic physics 154

S

spectral lines 77, 78, 83, 112
spectroscopy 55, 67
speed of light 61, 63, 64, 65, 83, 116, 122, 155, 156
sphere of fixed stars (last sphere) 18, 23, 30-32, 34 35 36, 38, 57, 59, 90, 97, 103, 104, 109, 153
standard candle 56, 81, 138, 157
standard model 75, 85, 86
star cluster 44, 49, 51, 55, 70, 73, 131, 135, 157

Stefan-Boltzmann law 126
stoic model (of the universe) 25, 26, 29, 39, 41, 52, 55, 62
subnuclear (subatomic) particle 79, 86, 87
supergiant 10, 71

T

thermodynamic laws 20, 48, 53, 87, 141, 142
tidal forces 123

V

vacuum 28, 73, 87, 128, 129
variable stars 56, 114, 138, 157
visibility in the universe 44, 45, 65

W

Wien's law 126

Index of Personal Names

A

Alpher Ralph 88, 126
Aquinas Thomas 97
Arago François Jean Dominique 120
Augustinus Aurelius (St Augustine) 15, 74, 98, 99

B

Bacon Francis 121
Beer Wilhelm 147
Berkeley George 91
Bekenstein Jacob David 20
Bentley Richard 41
Bessel Friedrich 95
Bethe Hans 126
Bolyai János 154
Bondi Hermann 46, 81, 100, 127, 137
Bracciolini Poggio 33
Bradley James 156
Brahe Tycho 9, 30, 39, 100, 101, 143
Bruno Giordano 34, 40
Bunsen Robert Wilhelm Eberhard 112
Buonarotti Michelangelo 121

C

Cassini Giovanni Domenico 155
Chandrasekhar Subrahmanian 114, 115
Charlier Carl 50, 51, 102
Chéseaux Jean 44, 45, 102
Comenius (Komenský) John Amos 104, 106, 121
Copernicus (Kopernik) Nicolas 31, 34-36, 41, 94, 103-107
Cusa Nicholas (Cusanus) 33, 34, 106-108
Cyrus Reed Teed 130

D

Dee John 109
Descartes René 34, 121
Digges Thomas 1, 12, 36-39, 84, 109, 110
Doppler Christian 110-113

E

Eddington Arthur Stanley 113-116
Einstein Albert 75-77, 85, 114-117, 121, 134, 154, 163
Engels Friedrich 1, 64

F

Fowler Alfred 137
Friedman (Friedmann) Alexander 76, 85, 120, 121, 125
Fraunhofer Joseph 112

169

G

Galilei Galileo 1, 39, 105, 121-124, 146, 155
Gamow George 88, 125-127
Gauss Carl Friedrich 154
Gilbert William 145
Gold Thomas 81, 100, 127, 128, 137
Gore John Ellard 57
Guericke Otto von 3, 41, 128

H

Hale George 140
Halley Edmont 5, 43, 44, 46, 49, 52, 59, 83, 84, 129-132
Harrison Edward 60, 132, 133
Hegel Georg Wilhelm Friedrich 20
Helmholtz Hermann 48, 68
Herman Robert 88, 126
Herschel John 46, 50, 68, 120, 134, 135
Herschel William 135, 136
Hooke Robert 149
Hoyle Fred 81, 100, 127, 128, 137
Hubble Edwin 55, 77, 78, 85, 123, 138-140
Humason Milton 77, 138, 140
Huygens Christian 124, 155

J

Jansky Karel 47

K

Kelvin (Thomson William) 48, 57, 68, 141-142

Kepler Johannes vii, 1, 3, 4, 17, 31, 38-41, 49, 61, 83, 100-101, 107, 109, 142-146
Keynes John Maynard 150
Kirchhoff Gustav 112

L

Laertius Diogenes 21, 34
Leibniz Gottfried 27, 149
Lemaître Georges 76, 77
Lippershey Hans 122
Lobachevsky Nikolai 154
Lovelock James 152
Lucretius Carus Titus 27, 33, 43, 53, 118, 146-147
Luther Martin 106

M

Mädler Johann 58, 64, 83, 147
Mandelbrot Benoit 50, 148, 160
Mayer Robert 68
Medici Cosimo de 33
Michelson Albert 58
Milne Edward Arthur 23
Morley Edward 58

N

Newcomb Simon 57
Newton Isaac vii, 2, 41, 43, 105, 109, 129, 142, 148-150

O

Olbers Heinrich Wilhelm 1, 44, 46, 81, 100, 150

P

Peebles Jim 89
Poe Edgar Allan viii, 1, 60-64, 83, 152, 161
Ptolemy Claudius 25, 30, 153

R

Riemann Friedrich 79, 154, 155
Roll Peter G. 89
Rømer Ole Christensen 61, 122, 155, 156, 157
Rutherford Ernest 125

S

Schwarzschild Karl 121
Schwarzschild Martin 137
Seleucus 94
Shapley Harlow 55, 56, 141, 157, 158
Slipher Vesto 77
Smolin Lee 152, 161
Stukeley William 2
Symmes John Cleves Jr. 130

T

Tipler Frank J. 44, 60, 63, 161

V

Voltaire François 149

W

Wallenstein Albrecht 143, 144
Wilkinson David Todd 89
Wilson Robert Woodrow 88, 151, 158, 161

Y

Young Thomas 89, 58

Z

Zwicky Fritz 79

Made in the USA
Middletown, DE
31 December 2024

68559139R00113